PRAISE FOR OF

MW01493104

Not just for those who have served in our military but for everyone who aspires to be a leader and create a high performing team with a culture of excellence. The authors have masterfully captured not just the history of the Rangers who formed and led 2nd Ranger Battalion, but the timeless leadership lessons that echo today throughout the 75th Ranger Regiment.

—**Lieutenant General P.K "Ken" Keen,** US Army (Retired)
11th Colonel of the 75th Ranger Regiment

A magnificent book for any soldier or leader written by two seasoned soldier-leaders. Interesting and educational to read how values and traits were learned by individuals as they served in an early Army Ranger Company. You discover how they took these values to other Army units as well as their lives after military service. Special to me as my Ranger School training and service with a Vietnamese Ranger Battalion as a young officer helped set my values for my 37 years of service and the rest of my life.

—**Lieutenant General Freddy E. McFarren,** US Army (Retired)
Former Fifth United States Army Commander

This extraordinary book is filled with stories of patriotism, fidelity, selfless service, motivation, integrity, and a desire to be better every day at whatever you do. You will be immersed in a sense of loyalty and dedication to the Nation, the Army, the Rangers, and to each other. The original members of B Company 2nd Ranger Battalion came from many different backgrounds, experiences, and skill sets, but were brought together by one common goal – to be part of a unit that was committed to becoming the very best at what it did. This book gives the reader a peek into what it took to build one of the Army's premier light infantry units, whose legacy is now inscribed in the deeds and accomplishments of the thousands that followed the path laid down by these men and those like them.

—**CSM Jeff Mellinger,** US Army (Retired), 2nd Ranger Battalion Plank Holder and former CSM of U.S Army Materiel Command

This book is more than a compelling account of military history. It reveals the lifelong impact of modern Ranger training, and the impact disciplined, rigorous, values-based leadership had on these individuals, their families, and their unit.

It reveals the resiliency and the resolve, the toughness and persistence, the competence, courage, and integrity that shaped these soldiers throughout their careers and lives. Importantly, this book lifts up the lasting power that clear, effective vision and inspiring leadership can have on people and organizations. The individual stories allow all of us – whether we have served or not – to learn from each other, and to see ourselves. For this reason alone, *Of Their Own Accord* is a must read.

— **Brigadier General Anne F. Macdonald,** US Army (Retired),
President U.S. Army Women's Foundation

A powerful book filled with many accounts of courage, patriotism and humility. These soldiers embody the commitment to Selfless Service, putting the needs of their country and comrades above their own. The ethos of Selfless Service strengthens the moral fabric of our military, reinforcing the values of duty, honor, and country. It is this spirit of Selflessness that enables the military to function effectively, protecting the nation and its citizens with steadfast resolve. Thank you, Lawson and Fred for providing us with some needed inspiration in these challenging times.

— **Joseph D. Bray,** Civilian Aide to the Secretary of the Army

A display of exemplary leadership, often times under the most difficult of conditions. The quality of leadership shown here lead to lessons that we all can learn from. *Of Their Own Accord* should be required reading in all college level ROTC programs.

— **John Delavan Baines,** Founder & Chairman
Vietnam Veterans Memorial of San Antonio, Inc.

"We lucky few, we band of brothers." These words echoed in my mind as I read this powerful book. It tells the story of a single company of Rangers, highlighting the unbreakable bond they formed and the profound legacy they left behind. The foundation built in Bravo Company, 2nd Battalion, 75th Ranger Regiment, continues to benefit every Ranger today.

This book is the definitive account of modern Rangers, capturing their courage, brotherhood, and lasting influence. It's a must-read for anyone connected to or inspired by the Ranger legacy.

— **CSM Mike S. Burke,** US Army (Retired),
Founder of Legends of the 75th Podcast

OF THEIR OWN ACCORD

OF THEIR OWN ACCORD

A COMPANY OF ARMY RANGERS

CHANGING LIVES IN CHANGING TIMES

LIEUTENANT GENERAL LAWSON W. MAGRUDER III U.S. ARMY RETIRED

MASTER SERGEANT FRED R. KLEIBACKER III U.S. ARMY RETIRED

Printed in The United States of America

Design by Mark Babcock

ISBN 978-1-961505-27-8

Dedicated to the Original Members of the
2nd Battalion (Ranger) 75th Infantry

CONTENTS

FOREWORD

THIS BOOK IS BY SOLDIERS ABOUT A GROUP OF SOLDIERS WHO IN THE difficult days following the end of the Vietnam War volunteered to be a part of an experiment to rebuild the Army and restore the high standards of discipline and training that had become frayed after 10 years in Vietnam. The vision of then Chief of Staff of the Army Creighton Abrams was to create a unit that could become a model for the rest of the Army to emulate, and to have the veterans of that unit populate throughout the Army to pass along the best of what they learned from their experiences. To that end, the project has been a tremendous success as evidenced by the alumni of that unit who have gone on to serve in senior positions, both officer and enlisted, throughout the Army, and later as proud citizens of our nation.

I was blessed to be a young lieutenant in that unit, the 2nd Battalion (Ranger) 75th Infantry, when it was formed in 1975. Although this book is written by Rangers from another company, I shared common experiences with them during my two formative years in the battalion. It was there where many personal values were forged that have stayed with me all my life: trust, teamwork, service to others, hard work, training to standard, discipline, and lasting relationships. These values were fundamental to any success my teams and I had when I was a senior military leader and diplomat.

I served with many of my 2nd Ranger Battalion fellow officers and non-commissioned officers during my decades of military service. I marveled at the tremendous impact each of them had in the transformation of our beloved Army into the magnificent force that helped bring an end to the Cold War, allied victory on Operation Desert Storm, and defense of democracy during our war on terrorism. Everywhere my fellow 2nd Ranger Battalion "plank holders" served, they infused high standards and discipline and demonstrated caring, compassionate leadership. Their legacy lives on to this day in the younger generation of Rangers who are adhering to the Abrams Charter embodied in the Ranger Creed.

The themes in this book are instructive for many audiences: for youngsters who are looking for direction and purpose in their lives; for youth influencers who are looking for examples of personal values to share with those they support; and for military and civilian leaders who are looking for stories that reinforce team values for those they lead.

Joining the military has never been just about the pay or benefits. They are important, but more important is serving the Nation, being part of something greater than oneself, doing one's part as a citizen, and embracing a willingness to sacrifice in defense of our Nation. I believe our Nation, and our American traditional values to include military service are currently under attack. Patriotism and the desire to serve our nation is at an all-time low nationwide. Recruiting goals have been missed by all but the Marine Corps the last few years. Clearly there are new challenges and unfortunate circumstances that are impacting our nation's youth. This trend needs to be turned around, and I believe this book, written by veterans, can help reverse this trend and demonstrate that serving your nation is still a noble and fruitful endeavor.

I strongly commend this book to anyone interested in understanding the benefits of service to our nation. I encourage parents, faith leaders, coaches, military veterans, recruiters, political and community leaders, and other youth influencers to read this book.

As the title connotes, we all need to dedicate our lives to changing lives in these challenging, ever-changing times.

General John P. Abizaid, U.S. Army Retired
Former Commander of US Central Command and
U.S. Ambassador to the Kingdom of Saudi Arabia

INTRODUCTION

AUGUST, 2024 — SAN ANTONIO, TEXAS

"We few, we happy few, we band of brothers…"
—William Shakespeare, Henry V

THE INSPIRATION FOR THIS BOOK OCCURRED AT A REUNION OF THE ORIGinal members of Bravo Company, 2nd Ranger Battalion. We call ourselves "plank holders," a term that comes from the Navy's recognition of sailors who were present when their ship was being built. We helped build our company from its activation in October 1974 until its certification as combat ready, 14 months later. We had the privilege of transforming individual Rangers into a highly skilled, winning outfit founded on competence and trust.

It was our second reunion in four years and the second one in Reno where fellow Ranger Mark Lisi and his spouse Cindy live. Mark is a tremendous connector who has been instrumental in reuniting our fellow Rangers through the years. He set the condition for special gatherings in his wonderful city. Each time, our numbers have grown as we reconnect after decades apart. And each time, Fred Kleibacker and I have marveled at what our friends have done with their lives over the past 50 years. Not all stayed in the Army; many got out after their enlistment. All went on to live lives of significance, lives of service to others.

Fred and I dialogued at the recent gathering about a range of topics with three always surfacing to the forefront: how our company and its Rangers fulfilled the vision of then Chief of Staff of the Army General Creighton Abrams when he directed the formation of the two Ranger battalions in 1974; the immense recruiting challenges currently faced by

XIII

our military; and the intangible benefits of military service shared by our brothers.

* * *

Here is the backdrop for the book: When the U.S. military emerged from the dark days of the Vietnam War in 1973, the Army's senior leadership made key decisions to transform the Army. There were many rapid changes that occurred over the next decade: the change from a conscripted to a volunteer force; formalization and upgrading of NCO schooling; the transformation of how the Army trained; and the modernization of five key weapons systems. Another key decision was the formation of modern Ranger battalions that had been inactivated after World War II.

In late 1973, General Abrams wanted to repair an Army that had been overstretched with its standards frayed from long years of war, societal ills, and budget constraints. General Abrams, looking for a way to repair the institution, harkened back to the exploits of World War II Rangers. He wanted the Army to have the same high standards developed by William O. Darby, the first commander of the 1st Ranger Battalion who valiantly led it during the African and Italian campaigns and James Earl Rudder the commander of the 2nd Ranger Battalion that scaled the cliffs at Pointe du Hoc on D Day.

In late 1973, General Abrams directed the establishment of modern-day Ranger Battalions. First would be the 1st Ranger Battalion, activated on January 31, 1974, stationed on the east coast at Hunter Army Airfield near Savannah, Georgia. It would be followed by the 2nd Ranger Battalion, which was activated on October 1, 1974, on the west coast at Fort Lewis, Washington near the city of Tacoma. General Abrams issued clear guidance. It has become known as the Abrams Charter: "The battalion is to be an elite, light, and the most proficient infantry battalion in the world. A battalion that can do things with its hands and weapons better than anyone. The battalion will contain no "hoodlums or brigands" and if the battalion is formed from such persons, it will be disbanded. Wherever the battalion goes, it must be apparent that it is the best."

General Abrams expressly wanted the Ranger battalions to be the standard bearers for the rest of the Army. The formation of the battalions was the tool he used to affect the restoration of values and standards in the Army. This restoration would be key to the Army emerging from its post-Vietnam period of decline. The values and standards established in the Ranger Battalions would be spread to the rest of the Army with the assignments of its veterans to units throughout the force. Proudly, as you will read in this book, the initial plank holders in Bravo Company went the extra mile in fulfilling General Abrams Charter.

I was blessed to be selected by Lieutenant Colonel A. J. "Bo" Baker to be the first commander of Bravo Company, 2nd Battalion (Ranger), 75th Infantry. I, along with the other officers and senior noncommissioned officers of the battalion, were volunteers who were handpicked by LTC Baker for duty in the battalion. We each came from a wide variety of organizations in the Army. What was common to each of us was that we were "triple volunteers" having volunteered for airborne training, Ranger training, and service in the battalion. We each brought varying talents and experiences to the unit, but we each were physically fit, willing to learn, and highly motivated.

A brief description of the Ranger battalion's organization, and missions is needed for the reader to fully understand what the individual Rangers and their leaders endured in the early years of the 2nd Ranger Battalion:

The battalion consisted of 636 officers and soldiers. It was organized into a headquarters company and three Ranger companies, each commanded by a captain. The headquarters company consisted of the battalion commander and his staff and the support elements of the battalion (communications, medical, logistics, and food service). Each Ranger company had a headquarters section with the company commander, company executive officer, an artillery fire support officer, a first sergeant, and communications section, three rifle platoons led by a lieutenant with over forty soldiers each, and a weapons platoon with two 60mm mortars and three 90mm recoilless rifles.

The typical missions the battalion would be given once it was combat ready were direct-action operations, raids, personnel, and special equipment recovery in addition to conventional or special light-infantry operations. It would be inserted into the objective area by parachute, helicopter, fixed wing aircraft, or watercraft.

Bravo Company was activated on October 1, 1974, and leaders and soldiers started to arrive at Fort Lewis shortly thereafter. In January 1975, the leaders went to cadre training at Fort Benning for a month to be updated on the latest doctrine, tactics, and techniques being taught at the Infantry School. In late March, Bravo Company and the other Ranger companies started seven months of progressive training from squad, to platoon, to company level. As the commander of Bravo Company, I had responsibility, with assistance from my platoon leaders and platoon sergeants, to develop the training program for the unit. A building block approach was taken with key tasks with specific standards developed for each level of training.

Training over the next seven months was very intensive and physically demanding. Many weeks were spent in the field improving our fieldcraft and technical and tactical prowess. The tasks were continually being reinforced as new soldiers arrived weekly.

My challenge as the commander was to establish a foundation of trust and teamwork within the company. I had to create an environment by my personal example for winning, consistent behaviors to be inculcated throughout the unit. I was confident I could do so because I had the full support of my battalion commander and his staff; had combat experience as a platoon leader in Vietnam; and had commanded an airborne rifle company in the 82nd Airborne Division. I had also recently attended the Infantry Officers Career Course which was focused on company command and battalion staff officer duties. It also helped that we had experienced combat veteran senior noncommissioned officers in our platoons.

The ensuing months of training were extremely exciting and fulfilling.

With incredible support from LTC Bo Baker and his staff, Bravo Company was transformed into a competent, confident team prepared to be integrated into the battalion operations. Our skill was on full display in December 1975 during the battalion's certification exercise. During the challenging exercise, Bravo Rangers excelled on every mission and were recognized as the best company by the evaluation team.

This is a book of success stories. It is about Bravo Ranger "Plank Holders" who participated in the first year of training and were positively impacted by their time in the company and Army and went on to make a difference in our world throughout their lives. It is titled, *Of Their Own Accord: A Company of Army Rangers Changing Lives in Changing Times*. It refers to the 75th Ranger Regiment motto, "Sua Sponte" which is Latin for "of their own accord." The motto speaks to Rangers' ability to accomplish tasks with little to no prompting and to recognize that a Ranger volunteers three times: for the U.S. Army, Airborne School, and service in the 75th Ranger Regiment.

The chapters in this book are fifteen constant themes that came from 70 hours of zoom interviews with 40 Bravo Rangers. They are the intangible, not tangible, benefits derived from their service to our Nation. Common to all has been the value of Service to something greater than self which has continued long after their uniforms were last hung up.

Each chapter will be led off by me. I will share with you my definition of the behavior or theme and what I was trying to instill in our Rangers 50 years ago when I had the honor of being their commander.

In each chapter, we will focus on specific Rangers who reinforced the theme with their compelling stories. To learn more about these leaders, you will find their pictures in each chapter and short biographies after the Epilogue.

The intent of this book is not to recount war stories (of which we have more than a few), but to share the positive impact our time together many decades ago had in shaping how we lived our lives in the future. Our purpose is threefold:

1) Inspire young men and women to join the greatest military in the world.

2) Inspire the major influencers (parents, relatives, friends, veterans, educators, pastors) for America's youth to encourage them to serve.

3) Pass on a memoir from our fellow Rangers to their family members.

A special thanks to our "Band of Brothers" who inspired us and so many others through the years to "Be All We Can Be!" **Rangers Lead the Way!**

—LTG Lawson W. Magruder III, U.S. Army Retired

1. SERVICE

When you raise your right hand and recite your solemn oath of
enlistment to defend our Constitution, for many of you it is the first time
you are dedicating yourself to something greater than yourself. It is now
about "We" and not "I." You are going from "selfish" to "selfless."
—CPT Lawson Magruder

From left: Tom Gould, Darby Reid

MAY 1994, CLARKSVILLE, TENNESSEE: TOM GOULD WAS LOST, DRIFTING
afloat in the sea of regret. It had been six months since he impulsively decided to retire from the Army after 20 years of dedicated service and sacrifice. The loss of fellow Bravo Rangers and Delta operators Randy Shughart
and Grizz Martin and others in the Battle of Mogadishu had upset him
immensely. He strongly believed the denial of requests for support by the
senior leadership of our nation had contributed to the loss of his friends. It
had propelled him to retire. Now this tough Ranger was dejected, demoralized, and depressed. He missed the brotherhood, the friendships, the values, the focus on mission, the adrenaline rush, the sense of purpose, and yes,
the service to something greater than himself. Where would he ever find
"it" again? And then the call came from a friend, PJ Gorham, "Tom, they're

starting a Junior ROTC (JROTC) unit over at Lodge Grass High School on the Crow reservation in Montana. They need help. I think you're the right veteran to build the foundation. What about it?" That call set Tom back on azimuth and would refuel his purpose in life to serve others. This time it would be with a much younger generation that needed his leadership, experience, values, and warrior spirit to help them find direction in their young lives.

The consistent theme that came out of every one of our forty interviews was Tom Gould's purpose in life: selfless service or service to something greater than self. It was not surprising that twenty-nine out of forty of our fellow Rangers had been positively influenced by a father or an uncle or friend that had served in the military. Many had fathers and even mothers that had served in World War II and/or the Korean War. Some had relatives who had recently served in the Vietnam War. The eleven who had no relatives or friends who encouraged them to serve were self-motivated for many reasons. Some needed to get out of a messy family situation. Some wanted to find adventure and others direction in their lives. And some even wanted to break away from negative influences in their lives. Regardless of the reason for joining the military, they soon found out the value of service to the nation and to others.

Sadly, the desire to serve among young Americans has steadily declined in the last 20 years. Generational attitudes, and the culture wars going on throughout the United States, are highly likely affecting not only decisions made by today's youth, but also the advice being given to them by those who influence such decisions. The shift to a recruited-based army in which most of our peers entered had evolved over multiple years and multiple administrations—assisted by a Congress that had 80% of Senators and 74% of Representatives who had worn the uniform.

* * *

Unlike the period in which the professional, volunteer Army was created, however, 50 years later current cultural milieu as well as the lack of experience and understanding of the military among political leaders (only 34% of Senators and 18% of Representatives have served in the military) has adversely impacted recruitment into the military. As we mentioned in the Introduction, we would like to help the recruiting cause by highlighting the many positive benefits derived from service to our Nation. We want to lead off with the theme of Service by telling the stories of two servant leaders who have modeled it throughout their lives: Tom Gould and Darby Reid.

Tom Gould, from an incredibly young age, wanted to be involved in something that made this country better. Generations before him served in our military. He had ancestors who fought for the South in the Civil War. His grandfather fought in World War I and survived after being gassed. His father and seven uncles served in World War II. One uncle was in the Korean War. His father was in the Army Reserve until his passing in 1973. Simply put, Tom was steeped in military tradition throughout his formative years. When he came of age, it was a no-brainer for him to leave college where he was having great fun and continue the family tradition and enlist in the Army.

Tom enlisted in 1973 for the new Ranger enlisted option. It was a program that allowed young soldiers who had just completed advanced individual training to go to airborne school followed by Ranger School. Because young soldiers were extremely inexperienced, the graduation rate was extremely low. However, Tom successfully completed the challenging course the first time and was assigned to Bravo Company when it was activated. He spent over three years in the battalion followed by eight years in Special Forces and eight years in the elite 1st SFOD-D or Delta. His 21 years of service were characterized by a deep sense of duty, rigorous training, a commitment to excellence, and building strong relationships. Specific operations he conducted while in Delta will not be discussed for security reasons, but they required Tom to be physically, mentally, emotionally, and spiritually (purpose filled) tough. He had to endure and persevere many challenges that placed him in harm's way.

This quote captures the key lessons he learned as a senior NCO in Delta: "Having the knowledge of knowing what works, what doesn't work, and probably the idea of adapting to the situation of being able to switch horses midstream. I'd say if something didn't work, do it again or switch directions. As the old saying goes, the plan only goes till the first round is fired, and then it all falls apart. But you've got to be prepared for that and be able to adapt to the situation. And you've got to know your people, what their strong points are, their weak points, and who can do what and who can't do what. It was always interesting to see that some people would surprise the hell out of you and other people might disappoint you a bit."

Tom said his developmental years in Bravo Company taught him integrity, honesty, dependability, training to standard, and the importance of personal relationships. His years of service on active duty instilled in him a keen sense of duty and selfless service. He deeply appreciated the values and skills that he gained during his time in the Army.

Tom decided to retire in 1994 shortly after the Battle of Mogadishu where he lost several friends. He was disgruntled with the administration's senior leadership. He returned to Montana where service and giving back to others called him again. He volunteered and helped activate the first high school Junior ROTC program on the Crow Indian reservation in Montana. Junior ROTC instills in teenagers' confidence, discipline, hard work, pride in wearing the uniform, and being a part of a team.

Many youngsters who perhaps are not talented enough to play varsity sports find friends and a winning team in JROTC. Tom helped so many Native American children find direction in their lives with many going on to enlist and serve our Nation.

After two years teaching JROTC Tom was suddenly stricken by MS. This quote speaks to the personal courage, and resilience of Tom as he fought the disease: "When I was diagnosed with MS, I hadn't even made my first house payment yet. Wow. Me and my wife at the time had found

4

a house, bought it, and I screwed up. I didn't pay for the insurance where if you die or get disabled, a house is paid off. I didn't do that. I should have. And then for two years I was recovering from the disease and unable to work. I was hospitalized for a while, and I couldn't even feed myself. I would go to put a fork in my mouth and hit myself in the forehead. And I was laying there thinking, maybe if this is what I got to look forward to for the rest of my life, not being able to do anything, I ought to cap myself. That was the first time I really thought about suicide. And then I said, not knowing my luck, I'd miss and disable myself further, so I decided to fight. It took me two years to learn to walk again."

With typical Gould grit and resilience, he fought and won the battle against the disease and went into remission. He volunteered again to return to JROTC at a different school for another two years. During his JROTC instructor years, he inspired many disadvantaged youngsters to break free from their plight and join the military.

After our nation was attacked on 9/11, Tom was inspired to serve our nation again this time as a military contractor. For ten years, he served throughout the African continent training various allied soldiers in small unit tactics as part of the US counterterrorism and nation building programs.

* * *

Family also motivated Darby Reid to join the Army. In his own words: "On my mother's side, family goes all the way back to the Revolution. On my father's side, my grandfather, after he came over from Ireland, was in World War I. Then my aunts and uncles were in World War II. My father and uncles were also in the Korean War as well and then I had an uncle and cousin in Vietnam. After my dad got out of the Navy after the Korean War, he joined the California National Guard. He was in the Guard until they folded his unit and then he went into the Army reserve. And so, joining the military, just seemed like the thing to do."

Darby Reid's entire professional life was defined as service to others. Shortly after he graduated high school, he went down to the local recruiting station and volunteered for service in the airborne infantry. His entry on active duty was delayed for a year when he had to return home to help with the family tire business when his father was severely injured. After his father healed, he came on active duty and after airborne school volunteered for the 2nd Ranger Battalion. He split his four and half years in the Battalion in the weapons platoon and as the ammunition sergeant for the Battalion. He attended Ranger School after a year in the battalion.

Darby decided to depart the Army when his enlistment was up. He was enormously proud of his service but wanted to return to the family's tire business. When asked about what values and mindset he took with him from his time on active duty in Bravo Company and the 2nd Rangers, he responded: "So many of the values were just part and parcel of everyday life: integrity, respect, duty, personal courage. To me, I can see they just came out of everything else. If we were doing our job and doing it right, courage wasn't really an issue because everything else just fell into place the way we were trained, and in respecting each other. Having honor for the system, for the battalion, for the companies, for the platoons, for the gun crews was all never questioned."

What he cherished most about his active service was: the absolute commitment to excellence, the camaraderie, the mission focus, and the care and concern the chain of command had for its soldiers. These themes would serve him well throughout the next thirty years of his professional life as a soldier in the California National Guard and Army Reserve, and as a proud member of the San Francisco Police Department.

Darby enjoyed the physical nature of the tire business but wanted more excitement and fulfillment to continue to serve his nation and community. He soon found a Pathfinder position in a Guard unit while he "enlisted" into the police force. For 32 years he proudly served the citizens of San Francisco in a blue uniform. He held a variety of positions along the way: patrolman, sergeant, sniper, EMT, SWAT team member, and detective. He

OF THEIR OWN ACCORD

was a standard bearer as a trainer and leader. He was courageous in several dangerous situations. He would have continued to serve until a severe ankle injury ended his career. When he recounted his years of service in law enforcement, this quote reflects the positive impact his military service had on performance of his duties as a police officer: "As I learned in the military, the mission has priority. Get the job done, take care of the people around you, support each other. If somebody needs assistance, whether it's writing a report or is in a big fight and needs help now, it wasn't something that I had to think about. We just did it taking care of each other. Even if there were people that I didn't like in the police department that I thought were jerks, but that was fine when we're alone, or if it's just us. When it's us helping the public, then, yeah, that doesn't matter. Somebody calls for help, they need help. Doesn't matter. Doesn't matter what I think of a fellow officer as a person."

While his "day job" was as a policeman, Darby's part time job was as a soldier. In a variety of leadership positions in the Guard and the Army Reserve, he always sought the most challenging, adrenaline pumping positions. His journey as a part-time soldier was quite fulfilling. He served in a Special Forces unit, as a small boat specialist, senior NCO in a Signal and training units, and retired at the age of sixty as a the most senior NCO, a Command Sergeant Major in a Training Brigade. Throughout his service, he was focused on establishing and maintaining exacting standards and leading, coaching, and mentoring others.

But Darby's intense desire to serve did not end when he retired from the police department and Army Reserve. He put his ingenuity to use as a subcontractor for five years for an Air Force pararescue unit. He became the "go to guy" to repair their zodiac small boat pressure hoses, and their scuba gear, to refit all their military freefall oxygen systems, and rebuild their pre-jump pre-breathing panels. He thoroughly enjoyed serving the team while saving the government money.

For Tom Gould and Darby Reid, service to something greater than themselves gave meaning and purpose to their lives. For both, they con-

tinued a family tradition of service to our Nation. In their chosen professions they bettered themselves and those around them while they bettered
their community and the world around them. There are so many others we
could have spotlighted in this chapter, but Tom Gould and Darby Reid
best exemplified the Army's definition of the value of Selfless Service:

"Put the welfare of the nation, the Army, and your subordinates before
your own. Selfless service is larger than just one person. In serving your
country, you are doing your duty loyally without thought of recognition or
gain. The basic building block of selfless service is the commitment of each
team member to go a little further, endure a little longer, and look a little
closer to see how he or she can add to the effort."

2. ACCOUNTABILITY

"You must account for your own actions before you can hold others accountable. I use the concentric circle rule when things are not going well: I start at my own feet and then work out to determine who did not meet the standard. Corrective action then needs to take place immediately."

—CPT Lawson Magruder

From left: Danny Crow, Tom "Doc" Giblin

A RIFLE CAME FLYING OUT THE WINDOW.

Deputy Sheriff Danny Ray Crow had only been with the Tuolumne County Sheriff's Department for about six or seven weeks. It's a remote rural county encompassing over 2,200 square miles and home to about 24,000 citizens in 1983, about 11 people per square mile. The county is nestled in the Sierra Nevada Region of Eastern California and home to the northern half of Yosemite National Park.

Danny was on patrol in his 500 square mile area of responsibility when he was dispatched to a domestic violence call in the town of Twain Harte. The dispatcher instructed him that the suspect also had a gun. Danny and his Field Training Officer (FTO) sped to the location and arrived on the scene. As they were exiting their cruiser, a 30-30 hunting rifle unexpectedly came flying out of a front window of the house.

Danny assessed he might have an opportunity to get in quick and resolve this without violence, so his instincts took over. He and the other deputy pulled their batons and raced to the house where they announced themselves and made entry. There, they found a man and a woman inside the front room. The man was wearing a U.S. Army camouflage jungle fatigue jacket, the type typically worn by Rangers in that era (circa 1979). Standing in a classic karate stance with his hands raised and formed into human knife blades, it appeared he wanted to fight. As Danny was sizing up his opponent and deciding how to deescalate the situation, he noticed that the unit patch on the shoulder of the man's left sleeve was the coveted scroll of the 2nd Ranger Battalion.

Danny quickly assessed the situation and seized the psychological advantage by relaxing and calmly saying to the man "Sua Sponte." This is the motto of the Rangers. It's Latin for "Of Their Own Accord."

These two former Rangers, standing face to face, now both civilians, were just feet apart. Danny, a former Ranger NCO and Squad Leader, stood calmly at the ready to engage the man if he decided to get violent.

The man looked at Danny with a slightly confused look and said aggressively, "WHAT did you say?"

Danny peacefully replied, "Sua Sponte."

The man still looked confused but asked more evenly, "How do you know that?"

Danny steadily said, "I'm going to reach into my pocket and show you something." Danny reached in his pocket and pulled out his original 2nd Ranger Battalion "challenge coin" and flipped it to the man. The stunned former Ranger looked at the coin and dug into his own pocket and showed Danny his original 2nd Ranger Battalion challenge coin. Tensions eased as the man looked at the other deputy who was with Danny and firmly announced, "I'll go to jail with that guy," pointing at Danny as he turned around and placed his hands behind his back.

The two of them talked on the way to the Sheriff's office. Danny learned he had been one of the unit cooks when Danny was in B Company in the early days of the 2nd Ranger Battalion. They talked all the way back about who they both knew in common and reminisced about their time in the Battalion. While not what you might have expected, this is an example of personal accountability, but not from Danny. The man might have decided to fight, if not for the fact that he would have had to fight a fellow Ranger who knew many of the same people he did—people to whom he still felt accountable.

Danny Ray Crow always knew he wanted to be a soldier. He doesn't really know why, but since the age of eight he had always worn a uniform: Cub Scouts, Boy Scouts, and Police Cadets. He was further inspired to serve his country because of a cousin he really admired who was a Marine Corp veteran. Danny entered the U.S. Army in 1975, at the age of 18, two weeks out of high school. Danny said in his interview, "I felt a need to give back to my country because my country had given so much to me." Pretty rare and strong conviction for an 18-year-old kid.

Danny arrived at Bravo Company in October 1975. He was a 'straight leg' (non-Airborne) infantryman fresh out of Basic and Advanced Individual Infantry Training from Ft. Polk, LA. This made it all that much harder on Danny since the Rangers were an Airborne outfit. Arriving as a leg in an Airborne Infantry unit meant Danny spent a lot of time doing pushups to prepare him for Airborne School, and because of his "lower caste" in Airborne society. He couldn't get to Airborne school soon enough. He also had to figure out how to fit into a unit that had been filling its ranks since October 1974 and had been training hard since January 1975. By October, the company had been training six days a week doing intense combat training. Training had started with advanced individual infantryman skills, squad tactics, platoon tactics, and now as Danny arrived, Company operations.

In October 1975, the physical and tactical training was grueling and relentless, so the Battalion saw high numbers of young men quitting. The schedule was typically to parachute jump into the training area at zero-dark thirty, Monday morning. Then the company trained day and night for the next five or six days. The week of training typically culminated with a grueling 20-30 mile forced march back to the company area.

This was the life for the first modern-day Rangers, the new Spartans. This was how Rangers were made in the beginning. Melded through high standards in all aspects of their lives and forged with realistic and grueling physical challenges — those who stood the test, who didn't quit or get injured, remained because they were disciplined and accountable to themselves, but more importantly, to the man to their left or right. All of them were harder than woodpecker lips.

The grueling training had slimmed our lower enlisted ranks by 5-10%, so we needed privates. Danny was one of many new replacement privates arriving almost every day. He and the others had successfully completed Basic and Advanced Individual Infantry training (and some had finished Airborne school, enroute). However, that training could not compare to the rigorous training, with the necessary high standards, it took to stand up the new, modern Ranger Battalions. The battalion's training philosophy was for every soldier to learn to do his job when environmental conditions were the worst. So, we practiced being cold, wet, tired and hungry and learned to ignore discomfort so it would not distract us from our mission when the real shooting started.

This is where the real learning started — and it started with personal accountability. Danny's first test came immediately. He and a half dozen or so leg privates from the company were being sent to Airborne School at Ft. Benning, GA. They stood in a small formation with their Platoon Sergeant, Sergeant First Class Roy D. Smith, awaiting their transportation to Seattle-Tacoma International Airport to catch their flight. Sgt Smith was a highly decorated Vietnam veteran, who sent them off with six simple words, "If you fail, don't come back." Danny's first test of personal accountability had begun. Danny successfully passed Airborne School and the next year attended and successfully graduated Ranger School.

During his three years in Bravo Company, Danny had various duty positions and served in third and second platoons. He eventually became a Squad Leader and achieved the rank of Sergeant/E5. Danny said this was a significant turning point during his time in Bravo Company. He took this promotion very humbly and seriously. He was now a NCO and responsible for other people. He was eager to pass down the knowledge and values he gained from his mentors to those he now led. Loyalty to duty (personal accountability), respect, selfless service, honor, integrity, and courage were values that were instilled in him by the NCOs and officers under whom he served.

Overall, Danny's time in Bravo Company Rangers shaped him into an individual of strong character developed through the shared hardships with his mates, NCOs, and officers. One of the most important themes from those years that stuck with Danny to this day is not letting anyone down, not being the "weak link." That is the essence of personal accountability.

Danny left the Army in October of 1978 and attended the police academy. After graduation, followed by ten weeks of field training, he was a fully trained deputy. He commented on how he took his Ranger experience and training and applied them to his law enforcement job. Danny said, "I mean, I took everything from the immediate action drills we used, and I just modified them to fit law enforcement tactics. Same with the planning stuff, I modified that too, and I just turned it all around to fit law enforcement and create standard operating procedures."

Danny went on to serve 33 years as a deputy sheriff. In addition to his normal patrol duties, he served as a new deputy training officer for 20 years, training hundreds of new officers over his time. All the fully trained deputies patrolled solo, so it was a big responsibility to ensure they were trained right. Again, this is where personal accountability comes into play. He also served on both the county SWAT and search and rescue teams during the same 20-year timeframe. The drive to never let anybody down was very strong in Danny.

After 20 years of chasing bad guys, training new officers, kitting up with his fellow SWAT officers for high-risk search warrants or hostage standoffs, time had taken a toll on Danny's body. He knew it was just a matter of time till no matter how hard he tried, he would become that weak link during a situation. He made his decision to transfer out of being a beat patrol officer and finish his law enforcement career as the deputy coroner. This is the core of personal accountability, knowing when it is time to move on before you're responsible for getting someone hurt.

In the Rangers, accountability fosters trust among soldiers, as does open communication with superiors without fear of retribution. Everyone had a job to do, and you were expected to do that job to the standard every day, week in and week out. This included even if you were the only one who witnessed what you were doing. This required trust, competence, skill, discipline, and a good attitude.

* * *

Good leaders prioritized taking care of their troops and lead by example. One such example of this was Thomas Giblin, AKA "Doc" Giblin.

It was July 1977, the rainy (Monsoon) season in the Republic of Panama. During the day, temperatures were about 90° and the humidity hovered between 90% and 100%. At night the temperatures plummeted to below 70°. The entire battalion was finishing a grueling 3-week jungle training program at the infamous Jungle Operations Training Course (JOTC) at Ft. Sherman. Ft. Sherman is located at Toro Point, on the Caribbean (northern) end of the Panama Canal, on the western bank of the Canal near the Provincial capital of Colón. Ft. Sherman was built during the construction of the Canal to protect the approaches to the Canal and defend it from any attack by land, air, or sea.

This was our second year attending JOTC training. It was focused on two areas, individual skills and operational tactics. Individual skills included land navigation in the jungle, tracking, living (surviving) in the jungle, and identifying key flora and fauna (especially the stuff that could hurt or kill you). We then moved into team, squad, platoon, and company level tactics and operations. The final week was an exercise pulling everything together. We endured three weeks of unrelenting heat during the day, monsoonal rains and fighting off hypothermia at night. We adjusted to living in soaking wet clothes and kit 24/7, surviving a myriad of insect bites, poisonous plants, and 4" Black Palm thorns. We survived parasites boring into our flesh, dozens of marauding pit-vipers (including the most dangerous snake in the Americas, the Fer-De-Lance), dozens of different species of Boa Constrictors, Panthers and Holler Monkeys. Simultaneously, we were patrolling, attacking, and ambushing. We did this for days and nights on end in slippery, steep mountainous terrain under triple canopy jungle, of course with little to no sleep. As the saying goes, "everything in the jungle can kill you, including the enemy."

We completed the formal portion of the program, proud Jungle Warriors. Nevertheless, of course we weren't done, that would have been too easy. We were Rangers. Yep, we're going to spend another week in the Jungle under worse conditions. We needed to fine tune what we learned and improve upon it to write our Standard Operating Procedures (SOP) for Jungle Warfare, the Ranger way. However, this proved to be a much more difficult operational scenario. The terrain was so challenging that we had to medivac Rangers frequently during the exercise.

We infiltrated from the sea via Landing Craft Medium (LCM) troop carriers, somewhere along the remote Mosquito Gulf in Northern Panama. We embarked on an aggressive mission, four days and nights, conducting attacks on various objectives then moving all night to attack another objective at dawn and then repeat. We walked many, many miles crisscrossing steep, slippery mountains under triple canopy jungle, so dense that it was almost impossible to traverse day or night and so dark at night you couldn't see your hand in front of your face.

We accomplished the final mission and were ready to exfiltrate, clean our kit and ourselves and head downtown to Colon for a beer. Word came down the line that there were no helicopters for exfiltration, and we were walking out. Estimated distance 50+ miles and we were doing it as a Battalion over the next 24 hours. Six hundred men in a single file, moving day and night without stopping except for short breaks every hour or so, where we got some relief from the 70 pounds of gear we'd been humping for weeks. Bravo Company was the last in the long line. Doc Giblin was the only company medic we had left, so he stayed near the Company Commander. 3rd Squad was the tail end Charlie—last element and rear security. The column came to a stop for a short break. The squad of 10 Rangers were spread out about 20 meters long with the rear of the squad resting up near the top of a hill above the rest of the platoon.

A young, inexperienced Ranger in 2nd Squad was an M203 gunner. An M203 is an M16 with a 40mm grenade launcher mounted under the weapons barrel. It was the most powerful organic weapon in a squad during that era. He was placed behind his squad leader so the squad leader could direct his fire if they were ambushed. Shortly after stopping, there was an unmistakable "Thump" of a round fired from a M203. The young Ranger said audibly, "Oh shit!" Everybody followed the arc of the flying simulation grenade. You couldn't miss it in the pitch black as the round spit intermittent sparks from its burning charge and headed up the hill towards the end of 3rd Squad.

BOOM! Then dead silence for 20 seconds before someone shouted, "SEND DOC BACK! SEND DOC BACK!"

Doc grabbed his aide bag and quickly made his way back down along the trail to 3rd Squad where he rushed 100 meters up the hill to find another Ranger private sprawled out on the ground with blood all over the lower half of his body. The young private had been sitting, resting his rucksack against the hill when the round impacted between his legs, creating a fist size hole in one thigh and shrapnel over the rest of his legs. Three other Rangers surrounded him and were assessing his wounds and treating him for shock when Doc arrived. Doc quickly assessed the private to ensure there were no other injuries. He field-dressed the wounds and said we needed to get a MEDIVAC ASAP. Doc stayed with his patient until he was successfully evacuated by helicopter some hours later. Then moved with 3rd Squad to catch up to the main element.

In Doc's own words: "Probably the most interesting thing was when we were down in Panama and guys would come to me, and I was just amazed at some of the things that they had. It's like they'd be out in the field, and they'd have a great big pus ball on their arm or their leg or something, and I'd lance it and get all the puss out and put a bandage on. And then I remember one time in Panama. This is the only time anything bad ever happened. We were in Panama the second time, doing a night march, and we stopped for a short break, and in the back, I heard an explosion. And again, it was a company march, and I was the only medic. And from the rear, you could hear the message coming up, 'Send Doc back! Send Doc back!' So, I went back and what I found was a Ranger who had been basically wounded by friendly fire. It was pitch black. I couldn't see a thing. It had exploded right between his legs. So, I ended up treating the gentleman, and we got him out of there and back. And I guess that he did okay."

Author's note: The wounded Ranger was PFC Randy Shughart. Randy would go on in the Battalion and eventually try out for and be selected for the 1st Special Forces Operational Detachment – Delta in 1988. He would later be posthumously awarded the Nation's highest medal, the Medal of Honor, for his gallantry during the infamous Battle of Mogadishu during Operation Gothic Serpent in 1993. Better known as "Black Hawk Down" and made famous by Mark Boudin's book and a Hollywood blockbuster movie of the same name.

Thomas W. Giblin, Jr. hails from Binghamton, NY. He joined the Army after finishing college, inspired by his father who served in World War II and the Korean War. His father was wounded in Korea conducting "snatch" missions to capture Chinese soldiers behind enemy lines. He was decorated for valor and was also awarded the Purple Heart. Tom had a desire to be an electrician after graduating from college but was discouraged during an interview and instead decided to join the Army and serve in the hardest job the Army had. His recruiter said, "Do we have a deal for you! Airborne

Ranger!" Tom had no clue what an Airborne Ranger did, and when asked, his recruiter said, "Don't worry about it, you'll love it!"

Tom joined the Army with a Ranger enlisted option contract for the 2nd Battalion as a medic in 1975. After Basic he went to Ft. Sam Houston, TX for medical training and then on to Airborne School at Ft. Benning, GA. He arrived at Bravo Company as an Airborne medic. Tom completed Ranger School a year later and served as both a platoon and then company level medic during his three years at the Battalion. Tom took being a medic very seriously and focused on doing his job well, a model of personal accountability.

Tom summed up Ranger accountability saying, "I just found it hard to believe that we had such men that were so dedicated to the job and serving our country. I was just in awe of these men. I wish that I could have been as good as them… they'd be out there grinding away, getting that job done, making sure they didn't let anyone down."

Once again, we see how each Ranger devoted his time and energy to live the motto of the Army in those days, "Be all you can be," All of them did so, so they were not the "weak link" in their units. The overarching core principle, of being accountable when folks were watching or, more importantly, when they were not watching.

Tom left the Rangers and the Army in 1978 to seek his dream to become an electrician. Tom took the values he learned as a kid that were reinforced during his time in B Company, qualities such as loyalty, duty, respect, honor, integrity, personal courage, and being personally accountable for all your actions. He recently said when asked how his time serving in the Army, and the Rangers in particular, helped shape the rest of his life.

Tom said," One of the things that the battalion taught me was to never give up, even if you get knocked down, knocked on your butt, to get back up and keep on going. And that's one of the things that I've always tried to do, especially in civilian life and as an electrician, no matter how hard it gets, to keep on going and to persevere and do my job and get it done."

Tom's entire life was about getting the job done. He worked independently on most of his electrician jobs, so if it wasn't done right, it would have to be done over either by himself or someone else. And if not done right, he could face unemployment. That's accountability. Tom took his responsibility as an electrician as seriously as he took his job as a Ranger medic. It was his responsibility to get the job done and to do it right, period.

Today Tom is retired but keeps busy and healthy with another passion, hiking. He belongs to the Triple Cities Hiking Club where he leads groups on three to four hikes a week, some upwards of 10 miles. He regularly leads hikes to the summit of the highest mountains in the Catskills. His medical training, Ranger leadership skills and drive on attitude all come in handy making sure people are safe, treating minor physical ailments, helping to navigate and keeping up morale and leading by example. Once again — being accountable to those around him and teaching them to be accountable to others.

There's an old euphemism we used to say about folks who defined a certain special characteristic, in this case, "if you look up accountability in the dictionary, you'll see their pictures." That's a way to say these men are the definition of personal and unit accountability. They rose to the occasion again and again and strove to live by the Ranger Creed's unequivocal emphasis on accountability their whole lives:

"Never shall I fail my comrades. I will always keep myself mentally alert, physically strong and morally straight and I will shoulder more than my share of the task whatever it may be, one hundred percent and then some."

3. PHYSICAL & MORAL COURAGE

We all will have fears during periods of danger. Physical courage is
how we counter those fears. Moral courage is oftentimes harder to
demonstrate because of the higher risks to career, friendships, and the
like. But in the end, if you do the moral, ethical thing, you will sleep well
for the rest of your life.
— CPT Lawson Magruder

Courage is not ever feeling afraid. It is feeling afraid and going on, in
spite of it.
—Father Kevin Shanahan

From upper left: Eldon Bargewell, Tim "Griz" Martin, Randy Shughart

MAJOR GENERAL ELDON BARGEWELL HAD RECENTLY RETIRED AFTER
over 40 years of distinguished service and joined the Wexford Group, a defense contracting firm focused on providing battlefield observers for DoD

(Department of Defense). According to Colonel (Retired) Hank Kinnison, Eldon's longtime friend and a senior leader at Wexford, at one of Eldon's very first meetings with other former military consultants, he could not resist speaking up.

As related by Hank: "There was a multimillion-dollar proposal the group was reviewing for the Joint IED Defeat Organization. There was a recommendation that Wexford tie their consulting fee for the work they did, so they would get more fee the less casualties they had on their advisory teams. Eldon patiently listened to this recommendation, and immediately made it clear that what was being discussed was repugnant, and the idea of tying money to loss of life was completely wrong. The room went immediately quiet. Eldon had spoken and there was just dead calm after that. And so, they just dropped it like it was the wrong thing to do, which it was.

"He didn't ask if he should speak up. It was just the right thing to do. That showed how much respect everyone had for Eldon. As Dr. Mike Stewart observed, Eldon had a fine sense of when to speak and when not to speak. He tended to listen way more than he spoke, but when he spoke, it was for effect. And in that case, it was absolutely perfect."

This short moral courage vignette is only one example of the courage filled life that Eldon Bargewell lived until his tragic accidental death on April 29, 2019. In this chapter we will explore the stories of three of our fellow fallen Bravo Rangers who demonstrated the utmost courage in combat: Eldon who was awarded the Distinguished Service Cross in Vietnam, Randy Shugart posthumously awarded the Medal of Honor for his valor in the Battle of Mogadishu, and Tim "Griz" Martin posthumously awarded the Silver Star for his valor in the Battle of Mogadishu. All three of these Rangers reflect the Webster dictionary definition of Hero: "A person of distinguished courage or ability who is admired for his brave deeds and noble qualities."

Eldon Bargewell was born and raised in the fishing town of Hoquiam, Washington. He had an all-American childhood: loving, hard-working family, strong friendships, and a star football player. When he discovered that college was not for him, he enlisted in the Army for the airborne and special forces. He soon found himself in Vietnam assigned to the Military Assistance Command, Vietnam—Studies and Observations Group (MACV-SOG) a highly classified, multi-service United States special operations unit which conducted covert unconventional warfare operations prior to and during the Vietnam War.

It was so secret that its existence was denied by the U.S. government. The group reported directly to the Pentagon's Joint Chiefs of Staff, and much of its history and exploits were concealed for years from the public by a veil of secrecy and confidentiality. Eldon's unit operated cross-border and behind enemy lines. As the youngest member of his team, he soon was recognized for his absolute competence and leadership ability. His officer team leader was then lieutenant Dick Thompson. Dick shared the following about Eldon's courage in combat:

"I think it was his mindset. You've got to have the right mindset to be able to go out on those types of missions where what they do is they fly you out into hostile territory, they put you on the ground, they surround you with an enemy battalion or larger, and then they say, okay, go try to do your mission now and survive to get out. So, you knew it was going to be very difficult. Every time people were getting killed, just right and left, everybody was wounded. In fact, most people said if you just got wounded, it was a good mission because the alternative was you almost got killed.

"But there are people like Eldon who you feel it when you get around them. They're just different when they talk. People listen to them; people have respect for them.

"And even though he was a very low-ranking individual (Specialist 4th Class), the more senior NCOs and everybody listened to him. I mean, they wanted to talk to him. They had respect for him. And I could feel it when I was around him, how he listened to what I was saying. He had great ideas.

"And then finally we got an opportunity to go on a mission, and I got to see him in action, a lot of action on that first mission. And I thought, wow, this guy keeps it together when we're in heavy contact. He's keeping it together. He's fighting hard. He takes care of the people. And I was extremely impressed with him. He was a warrior, but he was going to take care of the people.

"We continued to go out together. Every time, he didn't freak out. He watched. He saw what was going on. We made decisions. We worked together as a team. And I really enjoyed working with him." Eldon would receive multiple Purple Hearts and valor awards in Vietnam.

Continuing the topic of personal courage, Dick Thompson had this to say about Eldon: "I think a lot of people have that kind of built into them, and you can see it when they're younger. You can see it when they play sports. You can see that aggressiveness. Go get it. And a lot of people have the courage to tell you your baby's ugly and really upset you. But Eldon could do it without upsetting you so much. Somehow, he got you to hear what he was saying and listen to him.

"So, I think that's a good place where he complimented me. When other team members started pushing back against what I was saying, he would step in and say it a little differently and help get my message through. He was much better at that than I was. Eldon got awards, but it was because he was trying to take care of his people, trying to take care of the team, and get the team out. It wasn't about, if I jump up and take out these twenty guys, I'll get a big award for it. That never entered his mind. That's just not who he was."

Dick Thompson went on to have a wonderful military career and after the military earned a PhD in Psychology. He is a preeminent expert on stress and is the author of *The Stress Effect: Why Smart Leaders Make Dumb Decisions—And What to Do About It.* Here is what he says about how Eldon managed stress and learned about leadership: "In general, everybody has a limit, and if you get pushed past that limit, then you can't function. And it's different for everyone. Some people could go out on one SOG mission and that was it. They would come back and tell you; I'm done. There's no way I'm going back out there again. This is crazy.

"And other people would just keep going and going, but might get to the point where they become addicted, and now they feel like they have to go, and they start volunteering to go with other teams just so they can get another mission in. But then there are people like Eldon, who was able to manage a lot of stress and dissipate it when he got back and kept going out on missions and some hairy missions and then came back and did some more.

"One of the things he told me, I don't know, a few years before he passed away, he said, I didn't understand in the beginning enough about leadership. And he said, after I got the crap shot out of me on a couple of missions, suddenly it dawned on me, leadership is different than what I was thinking, and I've got to focus more on leading than doing. But he said I had to get shot twice to finally come to that realization that you can't always be the guy standing up and leading the charge over the hill if you're the leader."

Based on his demonstrated leadership skill in combat, Eldon was encouraged to become an officer. After graduating as the Distinguished Honor Graduate of his Officer Candidate class, he was assigned as a platoon leader in B Company Rangers at Fort Lewis, Washington. After a year, the company was folded into the 2nd Ranger Battalion when it was activated in October 1974.

Eldon was the platoon leader for the 3d Platoon. In Lawson Magruder's remarks at the dedication of a memorial in May 2023 in honor of Eldon in his hometown of Hoquiam, he focused on four dominant attributes of Eldon: his competence, his humility, his caring nature, and his courage. "From the very first week I spent with Eldon, two attributes came to the forefront: his humility and his absolute competence. He quickly became the informal mentor for his fellow platoon leaders. He also became the role model for his NCOs and young Rangers. He has been revered for decades because of his humility, competence, and indomitable courage.

Let me bring to life his humble attributes with some testimonials provided by his peers and subordinates."

Colonel (Retired) Marshall Reed, a fellow platoon leader, stated: "Since we were both platoon leaders in Company B, we technically were peers. The word "Peer" appears in quotation marks to highlight the fact that there's no planet upon which I could be Eldon's "peer." The gulf between our respective life experience and soldier expertise was simply too great. Yet, despite this cavernous difference, he treated me as a peer, not as I was a rank amateur. Eldon's focus was always improving individual Ranger and unit effectiveness. He recognized that an important part of his role in 2-75 Ranger Battalion was as a role model for the other lieutenants, his "peers", as well as his subordinates. He did this without ego. In a quiet, positive, and supportive way, he helped us grow as Rangers and leaders, making us and Company B far more effective."

Major General (Retired) Jim Jackson, also a fellow platoon leader, said: "Eldon was the true example of the 'Quiet Professional.' He lived by the standards he expected all to follow, and he did so without fanfare or need for recognition. Given his past service and the decorations he received, he was very unassuming and never talked about his past or his accomplishments. His approach to leadership was straightforward and real... he expected success and did everything within his powers to make it happen. As a fellow 'young officer' being around him was both interesting and informative... All you had to do was watch and listen. Eldon personified the old EF Hutton ad... when Eldon spoke all should listen."

Brigadier General (Retired) Bill Leszczynski had this to say about Eldon's humility:

"Eldon was a very friendly person when he could have been quite arrogant and condescending based on his many accomplishments since he enlisted in the Army, and specifically during his combat tours in Vietnam. We all knew what he did to earn the DSC. However, I never heard him say a single time anything like, 'This is what I did in Vietnam.'"

From Sergeant Tom Gould who served with Eldon in Bravo Company and later in SFOD-D (Delta): "He always allowed me to do my job and to be an NCO. I think it was the fact that he didn't interfere with subordinates doing their jobs. He would tell you what needed to be done, point out things that he wanted done a certain way, then allow his subordinates to do their jobs and would assist them if they asked or saw a major problem arising."

From Ranger Kim Maxin, who was a young sergeant in his platoon: "He led by example and from the front. His expectations for himself and for the soldiers around him were exceptionally high. He was a genuine leader and person. He never pretended to be anything but himself. Quiet, thoughtful, knowledgeable, and truly cared for his soldiers."

From a former junior enlisted soldier in his platoon and now a retired colonel, Bob Williams had this to say about Eldon: "Eldon was an unassuming, quiet, professional. He was not loud, boisterous, or domineering. He always seemed focused, with purpose and intent. He set an example of excellence and challenged you to that excellence. "Excellence" was the standard. He was not a man of giant stature, but his leadership style had a giant impact. As a subordinate and young Ranger, I felt both inspired and challenged to be worthy and accepted."

Co-author Fred Kleibacker captures the essence of Eldon as a young officer: "He exemplified confidence and leadership. No matter how fearful we may have been, he never showed fear. It was the consistent demonstration of his understated personal courage, without drama or fanfare, which was real inspiration. He always led from the front. We knew he would never commit us to something purposely foolhardy, even if ordered to do so. Our belief was always certain—no matter the situation, in training or combat, his combat experience, tactical knowledge, personal courage, integrity and loyalty to us and the mission would be the reason we would prevail. He would not fail us, even if we failed him."

After Bravo Company, Eldon went on to have an incredible career in the special operations field. He flourished in Delta as a tactical then operational leader. Most of the operations in which he participated cannot be discussed because of their classified nature. But one that has been publicly written about is the rescue of American Kurt Muse during Operation Just Cause in Panama. Fred Kleibacker relates the following: "When Eldon's squadron rescued Kurt Muse, they had overloaded the helicopter on top of Modelo prison, and they had to kick an operator off. So, the operator got off, and then Eldon had another guy get off. They both had automatic weapons. And then Eldon got off to be with his men, and he then let all the helicopters go and said, "Don't forget us. Come back and get us. Right?" And they stayed up there and continued to fight. The three of them finally got off Modelo prison when another Blackhawk returned and pulled them off the roof. So again, another example of Eldon taking care of people, always focused on others."

Eldon was a true American hero. But he was also a man who deeply loved his wife Marian and his family. He inspired his two sons, Brent and Logan, to follow in his footsteps. They also courageously served our nation during this current Global War on Terrorism.

Former Delta Commander and Chief of Staff of the Army Pete Schoomaker said it best about Eldon: "For more than forty years, since our earliest days in the unit, I was privileged to both serve with Eldon and to know him as a close friend. He was one of the finest leaders of his generation, living all seven of the Army Values to the fullest: Loyalty, Duty, Respect, Selfless Service, Honor, Integrity, and Personal Courage. Without a doubt, Eldon was a warrior ... the very epitome of the Warrior Ethos: "Always place the mission first; Never Quit; Never Accept Defeat; and Never Leave a Fallen Comrade. Eldon loved his country and his hometown of Hoquiam."

* * *

Randy Shughart and Tim "Griz" Martin were young "plank holder" Rangers in Bravo Company. Like many others, they were "triple" volunteers for the Army, airborne, infantry, and 2nd Ranger Battalion. Nearly two decades before their heroic actions during the Battle of Mogadishu on 3 and 4 October 1993, the noble behaviors they demonstrated as senior noncommissioned officers during that fateful battle were observed by their peers and superiors when they were young Rangers. Both shared a common behavior: they were eager to learn and grow every day as they practiced their craft. They were different in personality and stature: Tim was called "Griz" because of a birth defect that left one side of his face somewhat grizzled. He was strong as an ox, hence he always wanted to carry the heaviest load in his platoon. He possessed an outgoing, fun-loving personality. Randy was extremely focused and intent to prove himself and accomplish any task thrown his way. He was small in stature but also physically fit.

Tom" Doc" Giblin, a good friend, and roommate of Griz said: "I joined 2nd Platoon, Bravo Company, in December 1975. The Platoon Sergeant had the radio telephone operator (RTO) and the medic (me) following the weapons squad. We were assigned to the weapons squad in garrison until we went out into the field. The weapons squad consisted of two M60 machine gunners. Tim was a machine gunner when I got there, and he carried that gun like it was an M16. He was an extremely strong man. Physically, there was very little that he couldn't do. For some reason, Tim and I became pretty good friends. We used to drink beer together and play cards with other members of the platoon. We were all young and after our time in the field, we would talk about ways to improve what we did in the field—for the next time.

When we both made E-5 (SGT), we moved off base with SGT Mitch Erickson, and had a grand old time together. The first time Tim demonstrated for me his courage was not in the field but on a special project. On one of our parachute jumps, Tim broke his leg. During his healing process, the company and battalion were doing some long-range planning. They put Tim in charge of doing a lot of this planning, thus freeing up the officers and NCOs so that they could continue to train us. Tim was nervous (maybe even scared, but he didn't show it) about doing this. I do know that he wanted to do a great job and didn't want to let anybody in the battalion down.

He told me that he had never, ever done anything like this in his life, where he had to plan and come up with a solution. He was more comfortable using his talents to physically solve the problem. Now, he had to use his mind. Tim's detailed planning resulted in a tremendous success, and he was rewarded with a well-deserved Army Commendation Medal. Some may ask, where is the courage that is involved here? No lives were saved, no bad guys were taken down. All he did was come up with a plan. For somebody that is used to doing that, it's no big deal. But for somebody to have to do it for the first time, to get up off their butt, to learn and then implement ideas for a plan, and then do a great job, says something about that individual. To me, it shows that he had the courage to take on a difficult, completely foreign job to himself, and to successfully complete it."

Jim Dubik, Griz's second company commander remembers: "His physical strength was a manifestation of his inner strength, being able to work through his physical disability. I don't know what to even call it, but in his own mind, ignore it and be what he had to be. That was the thing that, for me, jumped out about him."

Griz left Bravo Company in 1978 as a young NCO for the Special Forces Qualification Course. Upon graduation he volunteered for and successfully passed the Delta assessment program. He would spend over 15 years in the unit as an operator participating in many dangerous missions worldwide. COL (Retired) Lee Van Arsdale served with Griz for over four years in a squadron headquarters. Lee said this about Griz: "Griz was selected to be the master breacher based on his demonstrated abilities. Explosive breaching is a critical component of the unit's tactics and techniques, and obviously has to be done right. He had life and death responsibilities that he took to heart, and with a great sense of humor and ready laugh. He was a pleasure to be around."

Young Ranger Randy Shughart was best remembered by Jim Dubik: "He was a specialist when I first met him, and the thing that jumped out to me was his intensity. I observed him several times in training, and he was focused. There was no random kind of energy around him at all. And I remember saying to First Sergeant Pross, I said, 'Man, that Ranger's either wired tight or he's going to be something special in the future.'"

Jim Smith, one of Randy's NCOs had this memory of him: "Randy was of medium build, athletic but not overly athletic, and always pretty quiet. He was always serious, well mannered, calm, and one of the politest soldiers in our platoon. He looked up to all his leaders and he was literally surrounded by great NCO's and Officers assigned to our platoon that were well deserving of his admiration, like Lieutenants Eldon Bargewell, Jay Dodd, Staff Sergeant Jimmie D. Bynum, and many others. After the day's training, Randy liked to stand around with his Ranger buddies discussing what he had learned and chat about how they could improve.

"I recall during marksmanship training his demonstrated skill and a reputation of being a great shot. Considering Randy's courageous actions in Somalia, I'm reminded that you never know how someone will react until they are truly tested. Randy's lasting impact on me, well, I often tell educators and scouters as we provide opportunities and teach citizenship to youth, you never know who that young man will grow up to be!"

Ranger Mike Shane recalls Randy with five buddies taking a trip south to Mount Hood, Oregon. "Randy was known to be very serious and was having some difficulty relaxing. But when he had some Jack Daniels, he then engaged in making a snowman laughing like a little kid."

* * *

Randy took the same path as Griz Martin and went directly from Special Forces Qualification to Delta. He also grew up in the unit as an operator. He and Lee VanArsdale were in the same operational squadron and knew each other well. Lee said this of Randy: "Due to the extremely realistic level of training in Delta, all operators are expected to — and can be relied on — to demonstrate physical and moral courage on a daily basis. Randy and Griz were no exceptions. Because of this, going into combat with them was much like conducting training as usual. You knew that you could count on them no matter the circumstances."

Lee said this of both these heroes who made the ultimate sacrifice during the Battle of Mogadishu, "When I say that they were typical op-erators, that's the highest compliment I can give. They were both down to earth and humble as anyone you'd ever meet. I wasn't surprised by their courage in combat, just saddened to this day at their tragic deaths. I'm a better Soldier and person for having been in their company. Both were family men who had simple dreams for their lives after the Army, which makes their sacrifice that much more poignant."

After Doc Giblin left the Army when his enlistment ended, he stayed in touch with Griz. They spoke frequently over the phone. Each time Griz encouraged Doc to come back into the Army and join the Special Forces and Delta. But Doc had started a family and his own electrical business.

His words about his fellow Ranger is a lasting tribute to a great American hero: "Tim was a good man and a great soldier. The US lost a great soldier that day in Mogadishu, Somalia. Hard to believe that it's going to be 21 years. I've been down to Arlington National Cemetery on 3 October a number of times to visit his gravesite. The first time I went down, I was wearing my OD field jacket with my Ranger tab on it. There were a number of soldiers buried in the same section that also had died that same day, and their friends had come down and left keepsakes for them. I took my Ranger tab off and left it at his headstone. I figured that he deserved it a helluva lot more than I did. I have never replaced that tab. Every time I look at that field jacket, it's a remembrance of what my friend did for our country."

Ron Buffkin remembers Randy Shughart. "I was Shughart's squad leader (2nd Squad, 3rd Platoon) in late 1977 and nominated him to go to Ranger School. Many outsiders think all Rangers are big, brawny linebacker types and while we had those aplenty in B Company, Randy didn't look like that. Randy possessed a reserved confidence, almost a muted personality until you got to know him. None in our squad in 1977 would prophecy that 16 years later Randy's courage would result in the Medal of Honor and cost him his life, but I can report that none of us would have been surprised. Randy was that kind of Ranger and human being.

"We did a lot of squad-level training back then, and I remember our squad, including Shughart, climbing up the rappel tower at Fort Lewis to practice the 'Australian Rappel.' The Australian Rappel is where you're 60-

feet up and face away from your anchor point with the rappel rope behind you. To perform it correctly, you lean over the side at the top and run down the face of the tower. On that day, Randy would pull lots of slack out of the rope and then leap off the side of the tower yelling, "Ranger," then come back up and do it again and again. We had a couple of new guys in our squad and them seeing Shughart leap over the top motivated them and helped them be less afraid of the height.

"Another time, B Company went to Mount Hood in Oregon for training. We attacked a target and then evaded, by squad, up the mountain for extraction. A blinding snowstorm prevented the Chinooks from Fort Lewis picking us up. Since we were on the ground longer than planned, I remember Randy dividing his remaining C-Rations among the new guys. I went to him and asked why he did that since the Rangers were "newbies" and needed to learn of their own accord. Randy didn't care; he was going to take care of his fellow Rangers. That was a lesson on that mountain for me as well."

After the 1993 Battle of Mogadishu and Randy's posthumous Medal of Honor, the US Navy commissioned a supply ship (USNS Shughart AKR-295) named in his honor to memorialize his valor and courage. Lawson Magruder, when he was commanding general at Fort Johnson (then Fort Polk) and his Command Sergeant Major Art Cobb, in 1994, named the MOUT (urban training site) "Shughart Gordon" after the two heroes of the battle. A post office and street in Pennsylvania are also named after him, along with several other monuments, including one at Carlisle, Pennsylvania where he is buried. The authors like to think Randy would only give a shy grin at a ship carrying his namesake.

Rudyard Kipling has a quote attributed to military service, that goes, "the ore, the furnace, and the hammer are all that is needed for a sword." Randy's courage was forged in the Rangers, and he became a sword.

"Greater love has no one than this: to lay down one's life for one's friends."
John 15:13

4. HONOR AND INTEGRITY

Integrity has many ingredients to it: honesty, trustworthiness, loyalty and reliability. If you live a life of integrity, you will bring honor to yourself and your outfit.

—CPT Lawson Magruder

From left: Roy "Rip" Prine, Ricky McMullen

THE DOOR SLAMMED CLOSED BEHIND HIM AND LOCKED WITH A LOUD metallic clang. The interior metal door lock released with a jolt and the green light came on announcing it was clear to enter. He pushed open the door and entered the prison dormitory.

It was 1990 and Roy Irwin Prine, or Rip, had recently retired from the Army after 22 years. When he got out, he really didn't know what he wanted to do other than go home to Georgia, specifically Valdosta. He decided to sign on with the Florida Department of Corrections at a unit 30 miles away from Valdosta. After training he was working in a minimum-security dormitory with a giant of a man named John. John was his sergeant. They got along well, and Rip learned fast under his guidance. Their friendship grew and so did the trust. So much so, that John said he would only work with Rip in the department's new facility in Madison, Florida. The new facility was a high security facility housing some of the most violent criminals.

Rip said of that experience, "I worked confinement for over two years. That's the jail within the prison for the most dangerous prisoners. In fact, I was asked if I wanted to do that job. They normally wouldn't allow a rookie to do that, it's a dangerous job. But I had a big old sergeant, he was about six foot six, and we'd worked together in my short time there. John stated, 'When they open the confinement facility at Madison Correctional Institution, I want Rip in there with me. I don't want any of these other guys. He's the only officer we have around here that I trust.' John trusted me to have his back. And believe me, in that confinement facility, you needed to have your back covered."

Where does trust come from? It starts with integrity.

Rip was older than most of his fellow correction officers (CO). He was a seasoned, Special Forces (SF) Non-commissioned Officer (NCO) who had spent years working in Central America during the insurgencies in El Salvador and Honduras in the 1980's. For most of the 80's, SF were omni-present in the region, training young, largely uneducated conscripted soldiers. These local soldiers were fighting to save their countries from violent communist insurrections, which were supported by the Soviet Union in the East and China in the West. The American soldiers were following in the footsteps of SF legends who had served in the decades long Vietnam War. Rip was part of a large cadre of SF NCOs and Officers training El Salvadorian and Nicaraguan soldiers how to fight and defeat the cancers of socialism and communism. The two ideologies that had killed hundreds of millions of people across the globe. During his time both in the Battalion and SF, Rip learned that building rapport, trust and respect for your allies was paramount. To do so, you had to lead by example, do what you say you're going to do, and show respect for people of different cultures. Honor and Integrity was and is paramount. As a CO, it was no different.

The inmates in Madison County Correctional Institution in Florida weren't much different from the soldiers in Central America. Most of them had dropped out of school, gotten into crime or some other kind of trouble. For the majority, this was not their first time in jail. It was a revolving door and to many of them, a badge of honor. Because of the high recidivism, the inmates developed an informally organized prison culture. If they committed crimes while in prison, they found themselves in Rip's and John's confinement facility. The worst of the worst. Rip knew in "their" environment, it was about honor and loyalty to their "criminal band of brothers." Rip knew it was about earning their respect. Keeping your word. Building trust. So, he never lied to them, but also called them out when they didn't reciprocate, with the understanding that, if they violated the "man's rules" he wasn't afraid to use the "big stick." Something Rip never had to do. He learned through years of training folks in foreign countries, the ability to read people and their characters. It was a required skill for survival in life, let alone training troops in foreign countries or managing inmates in a prison.

So, Rip carried himself as he did in the service, straight up, confident, quiet and unassuming. He carried into the civilian world a secret that was typical of the "Quiet Professionals" of Army Special Operation Forces (SOF) operators. Rip said of his experience, "It was simple. I treated them as human beings, unlike most of the younger COs. I developed rapport and mutual respect with the inmates. It's not that it didn't get violent between inmates, but we kept an eye on the signs of impending violence and tried to deescalate quickly. I never had to use my baton. You must live your life as a life of integrity. Be honest and live up to your word. I did a total of more than five years between confinement (two and a half years) and transporting prisoners (three years) to and from medical appointments. However, this wasn't my dream job, so I had to move on."

Rip was 18 years old when he joined the Army in 1968. Inspired by his father, who was a 33-year Navy Submariner, who had instilled in him the values of love of country—Duty, Honor, Country. Rip volunteered for Airborne Infantry and SF. He wanted to serve with the best and go to Vietnam. Times were hard for the Green Berets who were integral to the fight in Vietnam. They were fighting a secret war against the communist, North Vietnamese who were trying to reunify with South Vietnam; the country had been split up after WWII. The Green Berets had some of the most daring and dangerous missions. In fact, the first Medal of Honor awarded in Vietnam was awarded to CPT Roger Donlan, an SF soldier. Casualties were high and they needed a steady stream of replacements.

Rip spent the next five years in several assignments within SF, with his dream of going to Vietnam never materializing—it just wasn't meant to be. He learned that there was a newly activated Ranger Battalion at Ft. Lewis, WA, looking for volunteers, especially experienced NCOs. Rip had recently been promoted to Staff Sergeant (SSG) so he decided to find out more about what was going on and requested to be interviewed. The new 2nd Ranger Battalion Commander, Lieutenant Colonel (LTC) A. J. Baker and his senior enlisted NCO, Command Sergeant Major (CSM) Walt Morgan, interviewed Rip and hired him on the spot.

Rip arrived at B Company, 2nd Ranger Battalion at Ft. Lewis, WA, in January of 1975. He was assigned to B Company as the Platoon Sergeant for Weapons Platoon. At this point in his short seven-year career, Rip knew very little about conventional light infantry, how to run a platoon, what a weapons platoon's mission was and how it supported the company. He had a lot to learn and felt intimidated by other Officers and NCOs, most of whom had the combat experience from Vietnam he so valued. While he had received a warm welcome, it was also becoming very clear to him during those early days that a critical eye was watching everything. Rip said, "Officers and NCOs who couldn't meet the extremely high standards of the Battalion were fired faster than it took to cook a hotdog."

Rip found himself one of the original four Platoon Sergeants of B Company with the least amount of leadership and combat experience. The Platoon Sergeants of the other three infantry platoons were all combat veterans either with SF or Ranger Long Range Reconnaissance Patrol

(LRRP) experience from Vietnam. One of the Platoon Leaders in the company was a bona fide hero from the infamous Military Assistance Command—Vietnam, Studies and Observation Group (MACV-SOG), a Central Intelligence Agency mission using SF for their secret wars in Laos, Cambodia and North Vietnam.

It was now early 1975, B Company and the rest of the Battalion leadership departed for Ft. Benning, GA, the home of the Infantry, to attend "Cadre" training conducted by the Infantry School. Cadre training was designed to get the Officers and NCOs current on the newest light infantry tactics, weapons platforms, anti-tank weapons, marksmanship, hand grenades, land navigation, communications and to begin to build the leadership team within each company. They completed training and returned to Ft. Lewis seven weeks later to start evaluating their new soldiers pouring into the Company. These men were eager, wet behind the ears, 17–20-year olds, who were untested mentally and physically, but were volunteering to serve something bigger than themselves—their country.

Rip simply summed up his three years in the Company as, "It was an honor to be in 2nd Battalion. To me, that's what it was. It was an honor to be there. I mean, a lot of long hours, a lot of working weekends. That was just the way it was. But we were just proud to be in the 2nd Battalion...I've had so many guys that were in my platoon that said, 'I learned so much from you.' I guess there's something in it, whether it's my attitude, whatever, but I saw guys come and go, and you remember, we'd lose individuals almost every day. I think during the first year the battalion lost over 100 guys. They couldn't cut it."

When Rip left the Battalion in January 1978, the men of his platoon gave him a plaque with the simple inscription, "SSG Roy (Rip) Prine, the best damn Platoon Sergeant in the Battalion. Rangers Lead the Way! From the men of Weapons Platoon, B Company."

Not much more needs to be said about that. Honor through Integrity.

After serving with the 2nd Ranger Battalion, he served another 12 years with the 7th and 10th Special Forces Groups. He was stationed in

Panama and Germany and did two years as a Reserve Officer Training Corps (ROTC) Instructor. After retirement as a Master Sergeant, he signed on with the Florida Department of Corrections for over five years. Then using his Veteran Education Benefit, (GI Bill), he graduated with a degree in journalism and worked as a reporter for the local paper in Valdosta, GA. He wrote stories about warriors past and present. Why? To preserve their legacy for the community and their families. Rip's beat was the local military at Moody Air Force Base. Rip's background and integrity in telling the stories from the lips of warriors resonated with the local community. He told their stories, the stories of WWII, Vietnam, Panama, the Gulf War, and of course Iraq and Afghanistan veterans.

One of his most memorable stories was about a Vietnam vet. In Rip's own words, "One day, I got a phone call from a guy, he was a Nam vet. He said to me, 'I've read your stories, and I think you have a handle on what military guys think. I was a tunnel rat in Vietnam.' When he said that, my ears perked up because I knew a couple of tunnel rats. I met with him. He was 100% disabled with Post Traumatic Stress Disorder (PTSD). From what he went through, I could understand why.

"Veterans Day was coming up. I told my editor, 'I want to do a regular, small magazine with some stories about veterans.' I had a World War II guy and B17 aviator who still had PTSD. There was another who had returned from Iraq. The editor looked at the stories and said you can do one, it wasn't a hard choice, I did the tunnel rat. The story I did on the tunnel rat was published on a Saturday. He went to church the following day and called to tell me, 'As I was leaving church everybody was lined up to shake my hand and thank me for my Vietnam service!' He told me that's never happened to him before. He said, his fellow parishioners said to him, 'We didn't know what you did.' He thanked me and said, 'Now my grandchildren know what I did.'"

This is the nucleus of integrity that leads to a life of honor.

Today Rip is retired and lives with his wife, and best friend, in rural Florida along with their two cats, Zorro and Hondo. He is an active member with the American Legion. Believe it or not, he's still serving, albeit informally. Usually the best kind of service. His mission today is listening to veterans struggling with PTSD tell their stories and helping them work out their emotions over time. It's the little things that matter. Rip always said being in the 2nd Ranger Battalion was an honor and that he felt honored to be a part of it. He also learned about respect in the Army, that honor and respect were closely linked. During his interview we coined his story the 'Ripple Effect.' He embodied the unconscious desire to always help others through honor and integrity.

* * *

When you've lived a life of service to your country, your community and your family and friends, something greater than yourself, it can be a hard adjustment finding new beginnings as fulfilling as the military. Frankly it can be daunting. It takes patience, perseverance, resilience, planning, good mentors and most importantly, finding your new mission and purpose in life. Such is the story of Ricky Joseph McMullen.

The phone rang on Ricky's desk.

"EST Simulator, Mr. McMullen, how can I help you." Ricky said in his low gravelly voice.

"Mr. McMullen? Ricky McMullen? First Sergeant, Charlie 2nd of 502nd Infantry in the 101st Airborne Division?

"A long time ago, yes. Who's this?" Ricky inquired.

"This is Beach, I wanted to thank you for what you did for me." Beach announced.

"Beach? I recall the name. What did I do for you?" Ricky asked.

"Well, you kicked me out of the Army on a Chapter 13." Beach flatly stated. (Chapter 13 — Administrative Separations: Involuntary Separation for unsatisfactory performance, misconduct or other circumstances).

Surprised and taken aback a bit, Ricky said slowly, "Well, I might have helped put the packet together. But it was the Company Commander who made the recommendation to the Battalion Commander. But regardless, I'm sorry."

"No, no, no, you misunderstand. I want to thank you for doing that!"

Ricky said slowly with a dose of Missourian skepticism, "You want to thank ME for kicking YOU out of the Army?"

"Yes! I realized after I got out that I was my problem. After 9-11, I was still able to join the National Guard, since I only had a General Discharge. I've been in for 17 years now. I'm an E7 (Sergeant First Class) and have deployed twice to Afghanistan. I love it! I learned so much from you. So, I just wanted to say thanks for helping me to get my head out of my butt."

"Well, I'm proud of you, Sergeant First Class Beach. I really am. I'm really glad our actions helped you to turn your life around and it sounds like it turned out very positively. That last part is all on you, young man. Sounds like you just had to grow up."

Honor and Integrity. Passing it on.

When Rip took over the Weapons Platoon in January of 1975, he had a few privates who had arrived around the same time. Ricky was one of Rip's new privates. Ricky is a fire hydrant of a man. Short and muscular in stature, he was the perfect choice for the infamous M67 90mm Anti-tank and Anti-Personnel Recoilless Rifle. It was a beast, weighing close to 40 pounds, unloaded. At four and a half feet long, it was like carrying a small log. It was a relic of the Korean and Vietnam Wars. It was the only serious anti-tank weapon available for a light infantry company at the time.

Only the toughest of the toughest soldiers were part of the 3-man gun crew. Gunner, Assistant Gunner (loader), and Ammo Bearer, all shared carrying the rifle, humping the 90mm rounds weighing around 10 pounds each. Simultaneously, they carried their own combat load bearing equipment (LBE), which typically weighed 20-30 pounds and the Alice Large Rucksacks that, depending on the mission, averaged 45-65+ pounds, along and their individual weapons. In all, they packed out at 125+ pounds per man in all seasons, all kinds of nasty weather, in the jungles, up mountains, through deserts, and the arctic. We all respected the '90' crews, because none of us wanted to be on a 90 crew!

Ricky was an Army brat. His father was an Airborne legend. As he describes his upbringing, "Dad was a disciplinarian. There was a right way and a wrong way. And the old concept that you're the oldest, you're responsible, you're in charge. What goes wrong is your fault. Yeah, we were kind of run like a military organization. You're in charge of police call (picking up trash), you're in charge of the grass cutting, and the inspection will follow when I get home. Not in a nasty way, in a loving way, but very stern." It isn't surprising that the Army wasn't Ricky's first choice after graduating from high school, but as he put it, "It's all I knew."

Ricky enlisted in 1974 for the Airborne Ranger enlisted option. He arrived at the Battalion in January 1975 after successfully completing the normal path: Basic Training, Advanced Infantry Training, and Airborne School. Ricky was there from the beginning, slogging with the rest of us, but humping considerably more weight with his three-man crew. His crew was attached to 3rd Platoon, which was considered one of the best platoons in the Company and perhaps, in the entire Battalion.

Ricky graduated from Ranger School in September of 1975, just eight months after arriving in the Battalion. He got back in time to participate in the greatest physical event of all our lives — the Battalion's Army Test and Evaluation Program's (ARTEP) Airborne Infantry Combat Certification. It was a seminal test for all the Rangers in 2-75, regardless of rank. For three weeks the Battalion jumped, rucked, attacked, raided and ambushed day and night. To do this, it marched and operated through blizzards, sub-freezing temperatures, relentless rain, sleet and snow. Humping hundreds of pounds, only on foot, over hundreds of miles across brutal and unforgiving terrain.

Cold and physical injuries began to take their toll on many of these hardcore Rangers. Just like a real combat event, except we were spared from being shot at. In the end, the Company had almost as many injured

Rangers as healthy ones, with the caveat that the men who made it all the way through weren't necessarily uninjured.

Ricky spent 18 months humping the 90mm in B Company before he transferred to 3rd Platoon as a Squad Leader. It was a short stay. He was the first to reenlist in 1977 and headed to an airborne company in the 172nd Light Airborne Infantry Brigade in Alaska. He didn't know it at the time, but he was embarking on accomplishing the vision the Chief of Staff of the Army, General Creighton Abrams, had set forth in forming the Ranger Battalions. He was one of hundreds of Rangers, both NCOs and Officers, returning or venturing out to, the regular Army. They each took the standards, professionalism, values and leadership attributes imbued in them during their years in Bravo Company and applied them to other non-Ranger Army units. These men would eventually help to transform and rebuild the finest fighting force in the entire world over the next 20+ years. While Ricky still wasn't sure if he wanted a career in the Army, he reenlisted for another three years to make up his mind and to be nearer to his family home and his father who was still on active duty in Alaska.

Ricky went on to serve his country for a total of 24 years. He served in the conventional infantry, air assault infantry and mechanized infantry, as a Drill Sergeant and as a Ranger Instructor. He rapidly ascended the ranks and found himself serving in every enlisted leadership position. Ultimately, he earned the most coveted of enlisted ranks, that of Command Sergeant Major (CSM). He said of those post Ranger years, "Where everybody had that vision of achieving that standard and going above and beyond, I kind of carried that with me, and I was fortunate enough in most of my organizations to help them reach higher standards, but never at the collective level that we had in Bravo Company."

Ricky embodied the Ranger can do mentality forged by our unique training, the mental toughness required to serve in the Rangers, and the positive reputation of Rangers in the Army. He applied these to the units in his future assignments. Always leading by example, he improved every unit he served in, leaving it better than when he arrived. During the late

70s and 80s, the Army faced the challenges of racial tensions, drug abuse, alcoholism and many leaders and troops lacking dedication, discipline and professionalism. Despite all of that, Ricky pushed his troops and built both pride and a strong sense of camaraderie and brotherhood among them, like he had in the Rangers. He took the secret sauce of the Rangers' exceptional training, their mindset and their high standards and one by one rebuilt the units he led.

Ricky didn't stay in the Army to transform it, but that became a large part of his mission for 21 years: build cohesive, disciplined and honorable units that could defend this country. General Abrams would have been proud of his beloved Rangers in the 2nd Ranger Battalion. We built leaders by setting standards, having integrity and serving honorably.

When asked what common traits were exhibited by leaders, both junior and senior, in Bravo Company, he said, "The maturity level for the positions that they held, the wisdom that they displayed. I never saw them get really frustrated. They always kept things under control in terms of how they dealt with people in any situation that may have occurred."

He added, "Competence was instilled from the get-go: Task, Conditions and Standards. It was the rule of law. We always had to endeavor to do the very best we could at the highest level. And I never served in another organization in my 24 years in the Army that collectively had the people that would meet those standards as an individual and as a collective team. I carried that with me to the extent that I would probably diagnose myself as having obsessive compulsive disorder (OCD), which I've been told more than once I have. But there was never another organization I'd ever served with, regardless of my seniority, that collectively could achieve those standards we had in Bravo Company."

In the end, Ricky summed it up this way, "It's just the integrity piece, especially what those above me were doing, it was what I tried to do with those below me."

Today Ricky and his wife Belynda live in Missouri. When he retired from the Army, he worked for the Missouri State Prison system for a short while. After that, he returned to work with soldiers at Ft. Leonard Wood, MO, as a civilian contractor overseeing training. He finished up as a government employee serving in a key position in the Directorate of Plans, Training, Mobilization and Security for Ft. Leonard Wood. He is very active in his church, a small struggling Catholic parish, keeping it viable. He also helps to raise his grandkids. His son and daughter are also serving. His Chief Warrant Officer son has had seven deployments in this current war on terrorism while his daughter assists soldiers and their families as a Registered Nurse Case Manager at a nearby military installation. His family continues the tradition of selfless service based on integrity and honor.

Rip and Ricky humbly radiate examples of Honor and Integrity and are shining examples of this stanza of the Ranger Creed:

"Recognizing that I volunteered as a Ranger, fully knowing the hazards of my chosen profession, I will always endeavor to uphold the prestige, honor, and high esprit de corps of the Rangers."

5. OPEN, HONEST & RESPECTFUL COMMUNICATION

The lifeblood of any winning team is its ability to respectfully communicate: to discuss and dialogue; to agree and disagree; to create a feedback rich environment. But once a decision is made, everyone must pull on the rope together.
—CPT Lawson Magruder

From upper left: Bill Block, Marshall Reed, Jim Schwitters

1970, TAY NINH PROVINCE, SOUTH VIETNAM. BILL BLOCK WAS BACK IN Vietnam this time as an advisor to a Vietnamese airborne battalion. He served here before but in a US unit, 2nd Battalion 501st, 101st Airborne Division. This time would be different. He would be immersed in a company of foreign soldiers. Although he had gone to courses at Fort Bragg to introduce him to the Vietnamese culture and military, and to also gain basic language skills, he would be out of his comfort zone. But then he met his counterpart, Captain Kahn.

Kahn was an experienced combat leader who was conversant in English. In their first session, he quizzed Bill on his technical and tactical competence and was soon convinced Bill knew his job. Before their first mission together, he invited Bill to his home to meet his family and to share a meal with them. It was a relaxing evening as Bill was embraced by the Kahn family. That personal touch sealed their relationship. From that day forward Bill and Kahn were battle buddies openly and honestly communicating with each other often under the most dangerous conditions. Bill's experience as an advisor reinforced the importance of open, honest, and respectful communication. That personal touch is what Bill brought to Bravo Company as its top enlisted soldier, First Sergeant.

One of the keys to our early success in Bravo Company was that we understood that for us to rapidly grow as leaders, and as a new organization, we needed to share openly, communicate candidly, and provide timely feedback. We were on a very tight schedule (nine months) to field new equipment, integrate new members, and establish and enforce demanding training standards to become combat ready. We could not waste any training time along the way. Leaders at every level needed to openly share lessons learned with one another. There could be no coveting of best practices. The company command team of Captain Magruder and First Sergeant Block set the example for the platoon leaders and the platoon sergeants. Weekly training meetings at every level were rich with dialogue and professional, respectful discussions.

Feedback from observations during the day in garrison and in the field was instantaneous and deliberate. On the spot corrections were encouraged at every level. If an officer or senior noncommissioned officer was not complying with command guidance or Army regulations or not meeting established standards, the youngest Ranger could make corrections on the spot. This feedback rich environment fueled by open, honest communication built exacting standards, spurred trust, and yielded incredible team-work. Three wonderful examples of open and honest communications were demonstrated by the oldest soldier in the company Bill Block, the young-est officer, Marshall Reed, and a not so young junior enlisted Ranger, Jim Schwitters.

Bill Block enlisted in the Army in 1955 at the height of the Cold War. As a youngster during World War II, he heard names like Eisenhower, Patton, Marshall, and MacArthur at the dinner table and over the radio. He was inspired by those who gallantly led soldiers to victory like Rang-er Battalion Commanders William Darby and James Earl Rudder. As a young teenager he aspired to be an Army Ranger.

As soon as he reached 18, he enlisted and started a fascinating 20-year journey as a soldier. A journey that would take him to Paris, France as a combat engineer, to multiple tours at Fort Benning now Fort Moore as an infantry noncommissioned officer, to Korea with duty on the demilitarized zone followed by multiple tours in Vietnam with formal schools inter-spersed before he ended up in Bravo Company. Each experience along the way developed Bill into an expert communicator with an ability to connect with others openly and honestly. Noteworthy, he spent many years as an instructor and teacher where communications were paramount.

After graduating Ranger school in 1960, he was on the physical train-ing committee and became an expert in combatives or hand-to-hand com-bat. He learned how to provide detailed instruction and demonstration to young soldiers to ensure they were able to make instantaneous decisions and use proper technique in defending themselves from an aggressor. It was here that he first understood the importance of asking for and pro-viding specific, constructive feedback. This assignment laid the foundation for his tremendous success as a senior noncommissioned officer in combat and in peacetime.

As a reconnaissance platoon sergeant in 2nd Battalion 501st Infantry, 101st Airborne Division during his first tour in Vietnam, he quickly un-derstood the life and death importance of clear, concise communications particularly under dangerous combat conditions. During his second tour that lasted 22 months, he was an advisor to a South Vietnamese Ranger Battalion. There he mastered how to communicate expertly and patiently with his counterpart. He left Vietnam with several awards for valor to include the Silver Star but equally important he left with an understanding that to build trust you need to work extremely hard at open, honest, and respectful communications.

As a senior Master Sergeant, Bill was selected to attend the highest-level NCO formal schooling, the Sergeants Major Academy. It was there where he made lifelong friends, learned more about the "big Army," and leading at senior levels. He also learned innovative approaches and refined his oral and listening skills.

When he arrived to help form a new company, he was a seasoned professional who understood his role was to teach, coach, and mentor others. He knew that there was no such thing as over communicating when the pace was extraordinarily high. Of course, in 1975 there was no email or texting to help accelerate communications. It was all done in person which was, quite frankly, important in building relationships, trust, and teamwork.

Bill lived his mantra that: "Strong leaders of high character can accelerate learning and set the example for others." He demonstrated this for others in so many ways, but his hallmark was how he treated others. He could be quite demanding but never demeaning. He spoke to others of all ranks the way he wanted to be spoken to; in a nonjudgmental, respectful manner. He never used profanity. He did not hold grudges. He emphasized that consistency was most important, and he wanted no ambiguity in how he approached his responsibilities. He emphasized to all of us that soldiers needed to understand the "why" of a task.

Bill Block is a man of the highest character who is a gifted communicator. He is a strong believer in the values of duty, honor, and integrity and he demonstrated them as a soldier and still today as a tremendous volunteer in the veteran community in his hometown of Columbus, Georgia. For over 20 years, 87-year-old Bill and two other retired Rangers have

48

represented the National Ranger Association at every Ranger Class and Ranger Regiment Assessment Program Class. They are out at Fort Moore at 3:30AM eleven times per year on the first day of each Ranger Class and then later in the week at the end of the 12-mile road march to inspire and cheer on the students. They then attend the graduation ceremony and present awards to the officer and enlisted honor graduates. Simply stated, Bill Block continues to positively motivate the younger generation of soldiers with his positive, infectious spirit.

* * *

Like many other Bravo Rangers, Marshall Reed followed in the footsteps of family heroes and proudly served our nation in uniform. His grandfather was an infantryman in World War I and his father was a combat engineer in Europe and the Pacific in World War II. He attended Stephen F. Austin University and flourished in Army ROTC. He attended Ranger School during the summer break between his sophomore and junior years and was awarded the coveted Ranger Tab which was quite an achievement for a cadet. Commissioned a Regular Army infantry officer, he would become an "infantryman for all seasons" serving in a variety of organizations during his 28-year career: mechanized infantry, Ranger, Special Forces, and Airborne, culminating in being the requirements manager for two command and control systems. Each one of these organizations placed a high demand on Marshall's ability to be open and honest in how he communicated with others, which is what he demonstrated as a young lieutenant in Bravo Company.

Marshall arrived at Bravo Company after spending a couple of years in a unit that had mediocre leaders, low standards, and poor morale. He wanted to be in an outfit that reflected the "Army of his imagination." He found it in Bravo. As a young "naïve" officer (his words), he joined an outfit with handpicked, high performing officers and NCOs.

Marshall learned on his very first day in Bravo Company when he met

1SG Bill Block that the unit would be the model of respect and discipline that he had been looking for. Here's a story Marshall related to Bill Block at our last reunion: "I came into the orderly room (company headquarters) and reported in as a new second lieutenant platoon leader. I walked in and you immediately stood up, and I kind of looked around wondering why you stood up, and you said, 'Sir, can I help you?' And I said, "I'm just reporting in. I need to report to the company commander.

"You shook my hand, welcomed me, and asked me to have a seat and told me, 'I'll let you know when Captain Magruder is ready to see you.' And shortly after that, Platoon Sergeant Smith came and knocked on the door. You told him, 'Come in.' He walked in and went immediately to parade rest. And you talked, he then did an about face and left.

"Captain Magruder then warmly welcomed me. The respect shown by you, Platoon Sergeant Smith, and Captain Magruder that first day set the positive tone for my time in Bravo Company. That's when I knew I was in the Army."

Physically and mentally strong, Marshall led the way for his platoon in the most excruciating physical events like demanding road marches, field exercises, and airborne operations. While he excelled in the physical demands of the job, he soon realized that he needed to rapidly accelerate his learning by asking questions, digesting feedback, and communicating clearly with his NCOs and Rangers. He knew he was highly competent in his field of mortar gunnery and control of indirect fires, but he needed to improve his tactical competence and leadership skills. He did so by exploiting his communication skills. Never a 'shrinking violet,' Marshall strongly believed that there is no such thing as a dumb question. Extremely articulate, he spoke with a combination of enthusiasm and a passion to improve himself and the organization. He never shied away from asking a question for clarification or to offer a suggestion. He always did so in a respectful and professional manner.

"As he stated to us: 'I never hesitated to be forthright in opinion or assessment, no matter what the boss or peers thought.' As much as anyone else in the company, Marshall fueled the feedback rich environment that Captain Magruder and ISG Block were seeking."

After three years in Bravo Company, Marshall went on to proudly serve in many important pressure cooker positions for 22 more years, retiring as a full colonel. His last job in Army command and control systems resulted in him spending the next 17 years as a consultant, trouble shooter, and the eyes and ears for the active-duty colonels responsible for developing and fielding those systems. This work took him throughout the Army, including 14 trips to Iraq and Afghanistan during the war. Everywhere he went, he provided unvarnished, honest feedback to senior leaders with one purpose in mind: to provide the finest support to our soldiers fighting on the ground.

Here are some noteworthy quotes from our interview with Marshall. They speak to the importance of personal behaviors, values, standards, and open communication:

On behaviors: "There was a framework of expected behavior. The Rangers in my platoon expected me to act in a certain way. I expected them to act in a certain way. I don't think either one of us thought about it. It's just there, this thing we're trying to do, we're trying to be good Rangers. We have missions we've been given. And the most effective way to accomplish them long term is to act in a certain way: be disciplined, show up with your best effort. Everybody is a stakeholder. Take for example, the young private who says, "Hey, sir, you have a sling rope hanging out of your boot." Sure, he was enjoying making an honest spot correction of the Lieutenant. But the importance is that the standards weren't mine, they were ours. And everybody had a controlling stake in enforcing and executing them."

On values: "You can have somebody teach you values. There's book learning and there's the genuine inculcation of knowledge or whatever into your soul. And so, I didn't need anybody in the battalion to teach me that integrity was important, and I thought I arrived there with it. But to observe integrity as a matter of routine helps reinforce your own set of values. It's not like you are the only one who should have to hold people to standard. It's easier to hold people to standard if you're not the only one doing it right. Within the Bravo Company family, I didn't think about it at the time, but in retrospect, the values, and behaviors we're talking about were a matter of *unconscious routine.*"

On serving something greater than yourself: "It was about the company and the mission, right? So, it's not about you. You're a part of something bigger than yourself and you owe that to the soldiers that follow you. I always sensed that within our company, like the five deadly sins lying, cheating, stealing, laziness, and being disloyal, there were sins of commission and sins of omission, and we were expected to act and to behave and to treat other people in a way that's consistent with these. And so, watching and participating in that as a young leader, that made a hell of an impression on me that I carried with me the rest of my life."

* * *

We could have written about Ranger Jim Schwitters in several chapters particularly: service to something greater than self, physical and moral courage, adaptability, and resilience, and building trust and teamwork. But we decided on the behavior that he mastered at a young age and modeled throughout his magnificent career: open and honest communications. Of all the Rangers we observed as a junior enlisted Ranger and many others admired when he became an officer and led our Army's most elite unit and then later a general officer, Jim was known for his humble, cerebral, analytical, and introverted nature but a very present leader. He was a thoughtful observer of others, and a patient listener, with masterful verbal and written skills. To understand how Jim became such a respected warrior leader, we must explore his personal story.

James Henry Schwitters was born into a large family in the lower Cascade country near Seattle, Washington. The family shortly moved near the town of Winnebago in northern Illinois. Jim stated: "I was one of five children on a large family farm. I grew up in an idyllic family with traditional Christian values. We were immersed in and lovingly taught that character mattered. We were taught the value of hard work and respect for others; so, I fully credit my folks for giving me a leg up on what really matters in life.

"Because our life revolved largely around family and farm, I went to college not knowing much about life outside our community nor what the broader world was about. I went to a rural high school and was somewhat socially inept growing up. I was mechanically inclined, so I applied to LeTourneau College (now University) in Longview, Texas. They specialized in engineering and technology, which fit well into what I thought I wanted to do in life. I succeeded there and got a degree in engineering.

"I applied to work for John Deere in one of their major production facilities in eastern Iowa and married my wife Rebecca shortly after. After two years in Dubuque, Iowa, I was transferred to a new Deere start-up facility, still in eastern Iowa. That involved a move to the very small town of Dixon and our first experience at home ownership. We came to enjoy small-town life with the clear sense of close community and neighborliness. We became close to our next-door neighbors, who were quite a bit older than us and happened to have a son in Special Forces at the time. We enjoyed their friendship, and through their son, got to know a little bit about military service, something that I had not seriously considered previously.

"It was around their dinner table that I first came to understand what Rangers and Special Forces were really about. This really piqued Rebecca and my interest in service. And so, after talking to a local recruiter, I looked at what the enlistment options were.

"I gradually became convinced that it was my duty as a citizen to serve our nation. John Deere had a very generous program wherein they would keep my job, or an equivalent, open for up to four years. So, I enlisted, fully planning to spend four years, and come right back to work that I enjoyed and in which I found fulfillment. We also liked the small-town atmosphere there in Iowa."

At the time, the Army offered enlistment to a Unit of Choice, so Jim chose the 2nd Ranger Battalion for their proximity to family still near Seattle. The only openings at the time were as a radio operator or cook. Jim wisely chose to be a communicator and that started his journey to the battalion. After basic training, advanced individual training where he was the Distinguished Honor Graduate, and airborne school, he reported to B Company in March 1976. He was immediately assigned as the radio telephone operator for the company commander, CPT Lawson Magruder who was in his last month as the commander. He spent the next 18 months as the communications chief for the company except when he attended Ranger School returning as a new sergeant (E5) in April 1977.

When asked about what he learned while working for three different company commanders, Jim stated: "They all had, in their own way, the innate ability to inspire confidence through knowing their trade and being able to inspire others in tangible and intangible ways. One of my very first experiences with Jim Dubik was serving as his company RTO. We were doing a battalion mass tactical airborne operation into Yakima Firing Center. It was reported to be very windy on the drop zone; making overall conditions very marginal for a battalion parachute jump on a small drop zone. It was my first mass tactical jump and I sensed most of us had considerable anxiety.

"No sooner than we got into the aircraft, with me seated right next to Captain Dubik, he was notified that we had a jump refusal on our aircraft. Though it was none of my business, in my naivety and insouciance, I asked CPT Dubik what was going on. I was struck by the very calm manner in which he directed the soldier to take off his parachute and be seated in the front of the airplane, to be taken care of later. The speed with which he confidently assessed an unexpected situation, calmly, and competently resolved it, certainly inspired a new, young Soldier's confidence.

"Jim Dubik and the other leaders I observed in B Company, easily de-

54

veloped respect, had a calming presence, sure tactical knowledge, and un-flappability. They also had the ability to anticipate problems, plan for them, and embodied a sense of shared experience. All of those things inspire confidence."

Jim McNeme, Jim's good friend in the company headquarters said this about him: "Jim got things done, not in a whirlwind frenetic way, but in a quiet, organized manner."

Jim Dubik remembers Jim Schwitters from his young enlisted days to later when their paths crossed in Bosnia and Iraq when Jim was a senior officer: "He was always a quiet, very competent professional. Serious, in-trospective, thoughtful—even as an enlisted Ranger, more so later in his career. He was always a pleasure to work with."

At about the time Jim returned from Ranger School, he was getting a little bit bored. Keeping up with the day to day was challenging as was the physical conditioning program but he didn't feel intellectually challenged, so in his own words "I was casting about mentally; I didn't do anything about it, but was thinking, okay, what do I want to do for the remaining two years of my enlistment? Not that I was dissatisfied or had anything set for the next two years, but was thinking, do I want other experiences before I return to civilian life? And that's when I got a call out of the blue from our friends at Fort Bragg."

That call would change Jim's life forever. A few months previously, he had been among three NCOs from the 2nd Ranger Battalion selected to attend phase two, communications training, of the Special Forces Qualifi-cation Course. It was a very interesting experience for Jim because he got to train on the military's most sophisticated communications equipment and received very thorough technical training led by some outstanding NCOs. Jim excelled in the course and gained the attention of some key leaders in the Army's special operations world.

That led to the surprise phone call five months later when he came out on a selection list to join the Army's newly formed 1st Special Forces De-tachment—Delta. After some resistance voiced by LTC Wayne Downing, then the 2nd Ranger Battalion commander, Jim left the battalion for Fort Bragg. When Jim reflected on his two years in B Company, he stated: "I think every one of the seven Army values were embodied and exemplified in the rank and file, as well as leaders to an exceptional degree. I venture to say that had I not had the experience and lived around the embodiment of those values, I'd have not chosen to stay in the army. Personal courage is something I had to learn. I was never remotely athletic or prone to do anything involving strength or personal physical courage. Being in the battalion certainly helped me with that."

When Jim joined Delta at Fort Bragg, he was assigned directly as a communicator without being interviewed or having to go through the same assessment the operators went through. They just wanted to hire the best technicians, and Jim was one of the very best in his field. He went there initially as a communicator in what they call their signal squadron. That only lasted about three months when Col Charlie Beckwith, the commander, decided to embed within each of the operational organizations several technical people who would develop an habitual training and working relationship with the operational squadrons so that if their expertise was needed in an operation, they'd be fully integrated and ready to go.

Jim, with three others, were offered the chance to go through the extremely challenging operator selection process. Of the three, Jim was the only one to pass. As a result, Col Beckwith asked Jim to become an operator, which he proudly accepted. Jim stated: "I went through the first operator training course and snuck in the back door, so to speak, not having met several of the entry operator prerequisites, such as time in grade and rank. So, at that point in time, I was a charter member and the junior man in the entire outfit. A large percentage of the other NCOs had Vietnam experience, so I had to prove myself as a bona fide, contributing member of the team. I can attribute any success I had in doing so to my foundational experience in Bravo Company. So that got me in the door, so to speak. "

Then Captain Pete Schoomaker's first contact with then SGT (E-5) Jim Schwitters was during the formation of Delta: "We underwent assessment and selection in January- February 1978 and the first operator train-ing course together as plank holders in that regard. The weather during our stress phase was snowy, wet, cold, and miserable and at one point we were huddled in the dark in some old broken-down GP tent with an inoperable stove as protection from a storm. While most of us were adding every layer of gear we had, SGT Schwitters took the initiative to totally disassemble, clean, adjust, and reassemble the carburetor and fuel system of the stove without uttering a word. Before long, it was glowing and I told myself, "Now there's a man to remember.""

A few years later, COL Beckwith worked some "magic" and shortly after Jim was promoted to staff sergeant, he received a direct appointment as a captain in the US Army Reserve. He was then quickly brought on active duty as a captain in the Infantry. That would start Jim's fascinating professional journey as a commissioned and general officer.

Over the next three years, Jim served proudly in the conventional Army. It was a positive, yet extraordinary learning experience. After the Infantry Officer Career Course, he was assigned to a mechanized infantry battalion in Germany. He was thrust into the thick of things, initially serving as the battalion maintenance officer, which he enjoyed immensely because he loved mechanics and understood maintenance principles. He soon as-sumed command of the headquarters company, the largest company in the battalion with over 300 soldiers and more vehicles than the rest of the battalion combined.

It was in that mechanized battalion where he truly grew to love com-manding soldiers. Jim's personal observations from his time in a conven-tional battalion: "It just gradually dawned on me that, yes, there is personal fulfillment and satisfaction in being able to shape an organization and of having an impact, not just in mission accomplishment, but in the lives of Soldiers. It isn't all about being the guy in charge. We've all seen leaders, I suspect, who love being in command, but they aren't motivated to be com-manders in the fullest respect. It took me a while to realize that I enjoyed being in a position of influence solely to have the ability to influence in a positive way."

As his tour in Germany was ending, Jim requested reassignment back to Delta to be a troop commander and was accepted. He spent the next 11 years in the unit with a break to attend Armed Force Staff College as a major. Upon return to the unit, he was a squadron commander. Three years later, Jim attended the Army War College as a lieutenant colonel.

While in the Army War College, Jim was selected for colonel and command of a Basic Training Brigade at Fort Jackson, South Carolina. Though not traditionally a coveted command like an operational or tactical brigade, Jim was still pleased to be returning to troops.

Each of Jim's two years in command at Jackson were different experiences because of the contrast in leadership styles of the commanding generals he worked for. In the first year, he worked for a CG who was focused on statistics (graduation rates, etc.) while the CG in the second year focused on the quality of training and the intangible qualities of the soldiers who left the training center for worldwide units.

That second year, he worked for a true mentor, then MG John Van Alstyne. Jim enjoyed the second year of command and was personally fulfilled working with and deliberately trying to develop battalion commanders and teach them about training entry level soldiers. He was empowered to use much of what he had learned in Delta about training to proficiency in intangible qualities rather than focus solely on relatively easily measured skills and abilities.

He found at Jackson that a true intellectual understanding of how the initial training process should and could work was lacking, which jumps a step forward to Jim returning there as the commanding general several years later and being in a position to much more directly influence how that worked.

LTG (Retired) John Van Alstyne truly captured the essence of Jim Schwitters: "To this day, when Jim Schwitters's name comes up in a conversation, I ask folks to think of the person they've met who personified their definition of "The Quiet Professional." Then, I would state unequivocally that Jim would probably make that person seem like the "Chatty

Kathy" dolls our daughters loved so much as little girls. He is very likely the humblest, most unassuming, most private person I've ever known.

"Jim was a basic training brigade commander during the time I had the honor to command at Ft. Jackson in 1997-98. For two weeks or so, after arrival, I thought Jim was deaf and mute. I would observe him at meetings and gatherings and note that he rarely said a word. I began to watch him closely. Fortunately for me, in a short while, I figured out that he was brilliant, listened very carefully, and thoroughly analyzed everything. I quickly learned that if I could track him down in his office, he would quietly and carefully provide his well-considered thoughts on any topic I needed to discuss. Frankly, I came to frequently defer decisions on major issues until I could talk with Jim in his office.

"It proved a bit challenging to find him in his office, though, as he was almost always out observing and talking, seemingly casually, with commanders, drill sergeants, and young trainees. I quickly ascertained that his battalion commanders and company commanders respected him without bounds and welcomed opportunities to talk with him. Jim was the epitome of the coach and trainer we hear much about, but rather seldom see. Later, as the TRADOC Deputy Commanding General for Initial Entry Training, I observed countless brigade commanders. Jim Schwitters was, hands down, the most effective."

After a highly successful brigade command, Jim returned to Delta to become its senior commander. He and his courageous operators conducted many dangerous missions in defense of our nation. When Jim harkens back to his many years in Delta, he had this to say about officers integrating into the organization: "It's a very interesting organization with talent out the wazoo. Intellect is only a small part of that talent, as well as extraordinarily high motivation. It takes a unique commissioned officer, particularly at the entry and mid-level, to figure out how to navigate where you've got NCOs, most of them senior, who know more about the business than you will likely ever know. You need to figure out your unique role and develop your style, and not everyone does it well.

"In fact, we've had a few unfortunate circumstances where leaders just couldn't make that shift and fit effectively into their unique role. The hardest parts for many are 'let me think before I open my mouth.' And, 'I'm not the smartest guy in the room.' This is really sage advice for any young leader joining a high performing organization."

Jim departed Delta for the last time in 2002 when he was promoted to brigadier general. For the next six years he would serve in positions of immense responsibility. For two years he was Director of Security for Central Command Headquarters with service in Tampa, Florida, and Iraq. He ensured the security of facilities and personnel in the CENTCOM Area of Responsibility (AOR). During that period, one of the Chiefs of Staff he worked for was then MG Robert Clark who said this about Jim Schwitters leadership and communications style: "Jim was a great teammate on the staff. Very smart, modest, humble, relatively quiet, calm under pressure, and exceptionally effective. Everybody liked and respected Jim. He is one of the most unique and interesting people I have ever known. Jim is an unusually gifted, patriotic American. I have immense respect, admiration, and appreciation for him."

Jim's next assignment had him in Iraq working for then LTG David Petraeus. Jim was the CG of the Coalition Military Assistance Training Team with the challenging responsibility to train the fledgling Iraqi military. During this year, Jim brought to bear his vast training experience in the Rangers, Delta, and the basic training center at Fort Jackson. He was the right senior leader to accelerate training of a foreign army while they were at war.

After a challenging year in Iraq, General Pete Schoomaker, the 35th Chief of Staff of the Army, wanted Jim back at Fort Jackson, this time as the commanding general. "His mandate was to revamp the entire process... content and length, increase the rigor and relevance to meet the actual demands of combat so that no soldier was a liability on the battlefield, regardless of MOS, and most importantly... to install the Warrior Ethos in every Soldier. Jim's efforts were instrumental to our growing the Army to 82 Brigade Combat Teams, and over 200 brigade size fires, sustainment, aviation, intelligence, engineer, and a plethora of other units... active, guard, and reserve. He never missed a beat, even though resources and Army status quo culture were continuous challenges. He subsequently applied these talents in theatre with indigenous forces. Jim is a hero in my book. Pure Quiet Competence."

This gave Jim free reign in transforming basic training. During his three years in command of the Army's largest basic training center, Jim refocused the drill sergeants on outcome-based training: training designed to achieve a set of mental and internal competencies rather than simply meeting a set of skills and physical requirements. The latter are definitely important but can be very misleading. Simply put, if you focus on intangible and internal qualities such as values, character, initiative, resolve, resilience, indomitably, etc., in the overall context of military skills and ability training, the latter fall easily into place.

Perhaps the hardest lesson Jim learned was convincing the training cadre of the benefit to the individual soldier, as well as to the larger enterprise, of an understanding of the "why" of something rather than just the "what."

Jim emphasized: "Soldiers want to be challenged and they want it to be tough, but they want it to be fair and they want it to have meaning and not just going through a bunch of motions and yelling. I get very turned off by what I would call false displays of discipline when we all know that discipline is doing what's right when no one's watching. And it should come as no surprise that, when external stimuli creating the displays of false discipline are removed, so goes true discipline."

Jim retired in 2008 as a Brigadier General after 33 years of monumental service as a junior enlisted Ranger, a Delta NCO and officer, and general officer. Everywhere he served, he left his organization far better than he found it.

In retirement, Jim was involved in some training consulting work in support of Army training initiatives, but his passion became working boat operations in the waters of Bristol Bay, Alaska for the past eleven summers.

This hardcore Ranger worked for his brother as the captain of a large river tugboat. He worked out of the port of Dillingham hauling construction equipment and large refrigerated reefer vans with salmon. With his life-long interest in all things mechanical, he enjoyed mastering an entirely new skill set while spending summers in the "Last Frontier" and being "captain of his own ship."

Because it was time, Jim retired from being a sailor in September 2023. He and Rebecca are enjoying his final retirement in central South Carolina. One thing is sure, Jim will continue to live his purpose in life: "To leave the world a better place within the sphere of my influence."

We believe the words of General Baron Von Steuben who helped General George Washington bring good order and discipline into the ranks of the Continental Army best describe Rangers Bill Block, Marshall Reed, and Jim Schwitters: "You say to your soldiers 'Do this and he doeth it,' but I am obliged to say, 'This is the reason that you ought to do that,' and then he does it."

6. ADAPTABILITY & RESILIENCE

Winners are continually learning and growing. They understand there will
be "potholes" along the way that will slow their progress. But it's their
ability to admit their mistakes and to bounce back from adversity that
will inspire respect from others.
— CPT Lawson Magruder.

From upper left: John Snape, Bill Waterhouse, John Funderburk

THE AERIAL RIFLE PLATOON'S MISSION WAS TO RESCUE ANY SURVIVORS
and recover the dead from L Company Rangers, a Long-Range Reconnaissance Patrol (LRRP). It had been ambushed the day before during infiltration into the A Shau Valley in South Vietnam. The five ship, UH-1H,
"Huey" slicks (troop transports) transporting the platoon landed a short
distance away from the original Landing Zone (LZ) where the LRRP
team had attempted to infiltrate. This is where the North Vietnamese

Army (NVA) downed both of the American aircraft with withering automatic small arms fire and B-40 Rocket Propelled Grenades (RPG). The LRRP team and their brave pilots who were transporting them into the Valley of Death, as the A Shau Valley was known, were shot down before they had even landed.

As the Aerial Rifle Platoon moved slowly and cautiously towards the Rangers last known location, they came upon the still smoldering medivac helicopter that had attempted to get the wounded crew and Rangers out, noting all crew members and soldiers were dead. What the platoon didn't know is that the NVA were waiting for them. As the first squad moved past the wreckage to secure the far side of the crash site, they walked into a classic L-shaped ambush. 10 men fell dead immediately including their platoon leader. The rest of the platoon fell back to better cover and form a defensive perimeter.

The fight was on, but they needed to get to the other bird that they discovered had slid down a hill into a ravine where there were wounded Rangers and crew. They had survived but needed immediate medical attention and extraction. The NVA had let them live to use them as live bait to lure in more helicopters and troops. The platoon needed to secure that site and treat the wounded.

SGT John Snape was asked by another Lieutenant, who was accompanying the platoon, to create a diversion in the kill zone so they could get to the other helicopter. They needed to draw the enemy away from the ravine. Technically, John was a Supply Clerk, the only job he had been trained to do when he joined the army but had never done in his four years on active duty. This was his third combat tour as an untrained infantryman. Yet, in those three years in combat, John had acquired enough knowledge, skills and confidence to be promoted to a squad leader and lead other infantryman into battle. Incredibly, learning to control his fear along the way, he had successfully adapted to the life and ways of a combat infantryman, and he was getting good at it.

John quickly got two M60 machine gunners to volunteer to crawl with him to a lone tree, located up the hill from the crash site, in the middle of the kill zone. It was a dozen yards away from the ravine, between the enemy's position in front of the platoon and the helicopter where all the dead were located. Moving to the tree would hopefully draw the enemy away from the platoon and ravine so the rest of the platoon could extract and treat the wounded.

John ordered the machine gunners to wrap hundreds of rounds of ammo around their guns while he filled up his helmet with hand grenades. He took point followed by the gunners in line behind him. This small group of intrepid men then low crawled on their bellies the dozen yards to the tree. When they arrived at the tree, John placed the machine gunners on either side of him facing outward towards the enemy's positions. He remained in the center. He ordered the gunners to place grazing fire (no higher than 3 feet above the ground) into the enemy's positions. On occasion, he directed fire into the trees above the enemy to dissuade them from employing snipers. All the while, John remained kneeling and throwing grenades as far out to the front of their position as he could.

It must be working, he thought, as bark and wood chips rained down on his helmetless head. They were taking increasingly heavy fire all around them as the enemy tried to take out the machine guns. After what seemed like seconds, but was more likely an hour, they ran out of ammo and grenades. They had to get the heck out of there. John led them back the same way they came, crawling all the way back to the platoon's defensive perimeter on their bellies. After ensuring his little team made it back safely, John entered the perimeter. The Lieutenant told John that the diversion had worked. The platoon had been successful rescuing the remaining LRRPs and the pilots. The pilots were badly injured but night was coming, so they would have to wait until morning to get extracted. The Lieutenant also told John reinforcements would arrive first thing in the morning to help them.

While not the preferable situation, the platoon reinforced their defensive perimeter and remained vigilant through the night fighting off enemy probes of their defensive positions. At first light, a company from the 101st Airborne Division air assaulted into another landing zone and arrived on foot to relieve the beleaguered and beat up platoon. After clearing the kill zone and ensuring all dead and equipment were accounted for, the platoon and company from the 101st extracted with all personnel, dead and wounded, 100% accounted for, and returned to base. Mission accomplished. This was John's last combat mission in Vietnam. He was 21 years old.

This chapter looks at two resounding characteristics of highly successful people; their ability to adapt to chaotic and ambiguous situations and the resolve and toughness to overcome and survive adversity and move past it. In this chapter we're going to highlight three highly adaptable and resilient Rangers: John Snape, John Funderburk, and Bill Waterhouse.

John Charles Snape was born in Toronto, Canada. His family moved near Oakland, California when he was a young child. John discussed his life, before joining the Army, "I lived in a very poor community. My family was extremely dysfunctional. I never met my father. My stepfather at the time was an active-duty sailor, serving on aircraft carriers, rarely ever home throughout my school years. When I was in high school, there was an incident for which I was blamed, but I didn't have anything to do with. I even had an alibi that cleared me. But it didn't seem to make a difference. That incident made me decide that my life was not going in a good direction. So, I decided high school wasn't a good place for me. I wasn't doing well. I didn't see any future in it. So, I went down and tried to join the Navy. And the Navy kicked me out of their office and told me to go home because I was only 16. So, I went down to the Army, and they were eager to take me. It was 1966 and the Vietnam War was really heating up. I'm sure the Army was having a hard time finding people to join, so anybody that had ten fingers and ten toes could get in."

John couldn't join the Army right away though, he had to wait until he was 17 and get his mom's permission. John's mom supported him, and he took the oath to defend our nation on his 17th birthday, January 1, 1967. He qualified to be a supply clerk. After finishing basic training and supply clerk school, John was sent to Germany and assigned to an Artillery Battalion. It turned out they didn't need a supply clerk. In fact, they didn't want him. They reassigned him to the communications section of the battalion to OJT (on the job training) as a Field Wireman. It turned out, the commo guys didn't need untrained Field Wiremen either, but what they did need were bodies for unit details.

John found himself on permanent details—cutting grass, picking up trash, guard duty, ammo detail, manual labor details, kitchen police (dishwashing and food prep) and so on. It was an utterly depressing year, and he had another year to go in his two-year hitch (enlistment). He was fed up. When he turned 18 in January 1968, he went down to see his First Sergeant, and said, "I want to go to Vietnam." The First Sergent asked him why? He said, "Because I'm not needed here. I feel useless. I joined the Army to learn something and do something, and I'm not doing anything here except picking up trash."

Later that year, John got orders for Vietnam. When he arrived, he was assigned to an infantry brigade as an infantryman. When John protested and told the Sergeant he was not infantry, the man's reply was simply, 'You are now!' He was handed the brand new M16 rifle and a jungle rucksack, neither of which he had ever seen before. He got about a week's worth of training on basic infantry skills, and then they shipped him off to his new infantry unit. John would spend the next three years in Vietnam, in three different infantry units. John would see a lot of combat and it is where he mastered adaptability and resilience. Over time he realized he was not in control of his environment or assignments. So, he made the best of it and began to grow and mature into the soldier he was to become.

This is where his story started, his final mission, of his third volunteer tour in Vietnam where he helped to successfully rescue his fellow Americans. Before John returned to the states, his First Sergeant told him he had been awarded the coveted Combat Infantryman's Badge (CIB). John recalls, "I told him I wasn't qualified for it because it wasn't my MOS (Military Occupational Skill) of 11B (Infantryman). He told me they had put me in to get a secondary MOS. Once he returned to the states, he volunteered to be a Drill Sergeant and spent three and a half years "on the trail." It's an age-old phrase coined by drill sergeants. It refers to the time a drill sergeant serves pushing troops through basic training. Its origins spur from the frontier days of the Old West when cowboys journeyed from California to Colorado driving cattle.

While on the trail, he was promoted to Staff Sergeant (E6). His wife, at the time, also joined the Army and they were assigned to Ft. Lewis, WA in 1974. He was assigned as a cadre NCO at the Ft. Lewis replacement depot. There he met all the newly reporting NCOs and enlisted men being assigned to the newly formed 2nd Ranger Battalion. He observed them for over a year and decided that was what he wanted to do. So, in late 1975 he volunteered for the 2nd Ranger Battalion.

As he told the 2nd Ranger Battalion Command Sergeant Major (CSM) Walt Morgan during his interview, when the CSM asked him, "Why do want to join?" John told him, "I want to learn how to be an infantryman, a good NCO. I want to learn how to be a squad leader. I want to learn how to be a platoon sergeant. I don't know how to be one. I want to learn from the best because I see the best coming through here. I see these NCOs. These are top notch guys, and I know I can learn from them because they've done all of this. A lot of them are Vietnam Veterans. So, I know I am going to learn from the best, and then I'll learn everything I need to do my job."

John was resilient, he was not afraid to admit he didn't know something. For his first six years in the Army, all he wanted was to acquire the skills he needed to be an effective leader, with the appropriate knowledge, skills and abilities to be an NCO in the infantry. However, he was about to have a real, eye-opening awakening embarking into a world he did not understand. A world he couldn't understand. Welcome to the 2nd Ranger Battalion as a non-Airborne ("Straight Leg") and a non-Ranger qualified senior Staff Sergeant. As CSM Morgan told him, "You're going to take a beating in this outfit. They're going to harass the hell out of you, being a leg." John knew that, but he also knew, he'd been through tougher stuff, he'd been through three years of combat. He also knew, this was the price of admission to learn how to be the best NCO he could be.

John would serve two years in the battalion. It was a very steep learning curve. When he arrived, they put him in Weapons Platoon and assigned SGT Ricky McMullen to show him the ropes. Ricky was John's junior in rank, but his superior in knowledge. John shut up, listened, asked questions, studied and learned. He successfully completed both Airborne School and Ranger School. When he returned from Ranger School, he was moved to 3rd Platoon as the Platoon Sergeant and eventually he also filled in as Platoon Leader for a short period.

In John's words, "I just thought maybe a Platoon Leader's position can't be that much of a difference from the Platoon Sergeant. I was in for a rude awakening. I learned that there is a big difference between a platoon sergeant (NCO) and platoon leader (Commissioned Officer) and the roles and responsibilities in them and the concept and the character to the whole thing. I had a hard time getting my head around it because I was intimidated by superior platoon leaders that never offered to help me or anything, but they were my competition. I had a hard time. I was extremely intimidated, and it stopped me from becoming better and becoming stronger. It took me about a month or so before I finally woke up. I finally figured out that I had nothing to be afraid of. The light bulb finally came on, so to speak. So, I finally got in tune with the job. My motivation was simple. It was to look out for the soldiers. I didn't want any more people dying on my watch."

Throughout John's remaining career, he would find himself in similar situations arriving at units as a senior NCO and being placed in officer positions. His years in Vietnam taught him how to adapt to his environment and his mental toughness and can-do spirit demonstrated resilient toughness that allowed him to succeed in every assignment. His time in Bravo Company was a pivotal point in John's career. When he arrived at Bravo Company, he was a total outsider with little experience in leading highly trained units.

He described his time in Bravo Company with affection, "The camaraderie of the NCOs was incredible. They stuck together no matter what.
They helped one another. At the same time there was a lot of challenging
one another, a lot of competition between them, but it never interfered
with the camaraderie. And I liked that. I liked the fact that I was able
to grasp everything that I was learning. I was heavily intimidated in the
beginning and became so proud that I finally overcame the fear. The light
bulb finally came on, and I finally figured out that I was as good as anybody there. It took a lot of hard work, a lot of studying at home, a lot of
observation, watching. I'm proud of the fact that I didn't let it overcome
me and I never quit.

"The sad part is that's not the way it was in the rest of the army during
this era. I found that out after leaving the Battalion the units I was assigned
to were not trained as well as the Rangers. Because of what I learned in
the company, the training, the other NCOs and officers, my leadership
improved and made me a better leader. I applied what I learned throughout the rest of my career to make units better. It was more than any of you
could have possibly known."

John went on to serve for 26 years and retired as a Command Sergeant
Major in 1993. In those remaining active-duty years and into retirement,
John leaned on his ability to adapt and remain resilient through a lot of
trials and tribulations. After retirement, John eventually found a job as
a veteran's representative with the California Employment Development
Department. In his words, "My job was to help veterans find work and to
help them with their barriers to employment. I did a lot of studying, talked
to schools and talked to employers. It was like being a used car salesman.
You basically go around, find veterans looking for work and find out what
their barriers were. Then you go out and talk to employers and see if you
can make a match. And I did that for 14 years."

When asked about his incredible adaptability and resiliency and how
the military helped him John replied, "Well, let's say tact, compassion, listening skills, passiveness, not taking things too seriously and learning to

just let things roll off your back. Over time the military gave me self-worth and taught me to care about my job no matter how much I hated it." When John took the job to help veterans, he said, "It was just about always caring, because you got to remember that you're talking about people's lives and their livelihoods, and they fear not finding another job. I experienced it when I got out and nobody would hire me. I couldn't figure it out. I had a hard time with it and so I understood it. I learned how to never quit. Keep pushing, keep looking for other avenues, looking for other ways to approach this to get the job done."

* * *

Adaptability and resiliency are born from many mothers. They are the human ways of overcoming difficult experiences and learning how to adjust and build the mental and physical toughness to endure with a positive attitude. Sometimes suffering through physical challenges is much easier than dealing with the loss of a fellow soldier, a loved one or the end of a relationship with a spouse. Success in life is overcoming these challenges and continuing with your life's mission.

The giant C-141 banked into a left-hand turn into the wind for its final descent into Dover AFB in Delaware. An American flag draped a steel casket secured to the rear ramp with ratchet straps connected to the floor. The lone escort was Special Forces Staff Sergeant John Funderburk. He sat close by, securely buckled into the red, webbed bench seats that stretched the length of each side and down the middle of the cavernous Air Force cargo aircraft. Dressed in his formal dress green uniform, with spit shined jump boots, medals and badges, holding his neatly folded Green Beret in his hand, he sat staring across the empty airplane trying to make sense of what had just happened a short 24 hours ago. Within the casket was the beloved founding commander of the 2nd Ranger Battalion and Special Forces legend, COL A.J. "Bo" Baker. He was also John's father-in-law.

John was escorting his remains back to the U.S. from Flint Kaserne,

Bad Tolz, Germany, home to the 1st Battalion, 10th Special Forces Group. In 1980, the Cold War was at its peak and 10th Special Forces Group was responsible for conducting covert operations in Soviet controlled East Germany and its Warsaw Pact aligned nations if war broke out. John had run into his father-in-law early in the morning just the day before to let him know that his team was conducting special medical training that afternoon and he was welcome to come observe.

John described it as if it was yesterday, "He was coming out of the sauna with Command Sergeant Major Morrell, when I ran into him. He had just had major knee surgery and was rehabbing it by swimming. I mentioned to him we were having a special medical training event later that day at 1500 in our team room, and I invited him to observe. This was early in the morning, probably right after PT or thereabouts. He said he would come down. So, I walked around the corner to go clean up and change clothes after PT.

"I hadn't walked more than 15 or 20 feet when I heard this loud noise, it sounded like somebody throwing about 225 pounds of raw liver down on the ground. I ran back around the corner and saw a medic from another team working on him. Bo was unconscious but the medic said he was still alive. They took him in an ambulance down to the clinic. I quickly got cleaned up and ran back to the house to let my wife Terry, Bo's daughter, know what happened.

"She was in a tizzy. I said, 'Look, everything's going to be all right.' She told me, 'NO, HE's DEAD!' Wow. Okay. Shit."

The C141 taxied down the ramp and stopped in front of a large hanger. The casket was moved off the ramp onto the deck of the aircraft. As the ramp opened, John could see an honor guard, chaplain, and some senior SF officers waiting. An eight-man element of SF NCOs in dress uniforms stood at attention and when ordered by the detail's Non-Commissioned Officer in Charge (NCOIC), marched into the plane and moved to the casket.

As they entered, they split evenly, four on each side of the casket with the NCOIC marching behind them giving commands. The NCOIC gave the command, and the detail faced the casket, lifted it, and then carried it off the plane as the flag detail rendered honors. The casket was placed under a roofed structure with 50+ other caskets that had been coming in from all over the world. These were servicemen and women and family members who had died from accidents, training incidents, suicide, and natural causes. Dover was the hub to process the remains of the fallen and then ship them to their final destination for burial. John was escorting COL Bakers' remains to his hometown of Searcy, Arkansas.

The next morning, John went back to the base to get ready for the next leg of the journey, escorting the casket to Searcy. When he arrived, there was a panic ensuing. The night before there had been a huge storm that had whipped through the casket holding area and tore off many of the tags tied to the handles with the name of the soldier inside. The escorts and staff now had the grim task of having to open caskets to identify their fallen comrades. It was traumatizing for everyone.

As John tells it, "We started at the top left-hand corner of the structure where the caskets were resting, and started opening caskets, looking for Colonel Baker. And it was just lucky we didn't have go through 50 or 60. I think it was probably about a half a dozen or thereabouts. But that was my introduction to death, to be honest with you. I mean, I've had some people die before, but having to open caskets and see the remains of soldiers who had died in awful accidents was traumatizing."

John Britt Funderburk grew up in the Panama Canal Zone from the age of six. He was raised by his single mother who was a WWII Army nurse and had moved them to Panama when she took a job with the Panama Canal Company. She worked as a nurse in the company's hospitals in the Canal Zone. Since he was a kid, John always wanted to be a soldier. He had a cousin who was assigned to an Operational Detachment in the 3rd Battalion, 7th Special Forces Group at Ft. Gulick, Panama, located on the Atlantic side of the Isthmus. This is where he and his mom also lived during his high school years. John had another cousin who he respected very much who was a veteran and a diesel mechanic. John decided after graduating he would attend diesel mechanic school like his cousin.

After he graduated in 1971, he left Panama to attend a diesel mechanic school in Arizona. After a few weeks, he decided the program wasn't for him, so he quit and worked at an ore mine in the area for a year. In 1972, he moved to Ft. Smith, AR to marry his high school sweetheart Terry and worked locally for a couple of years. After their first child was born in 1974, John decided it was time to do something with his life and joined the service.

He discussed his choices with his new father-in-law, Bo Baker, who wanted him to join the Army and go to OCS. John decided OCS would not be his best choice, so he enlisted for the Ranger Enlisted Option that year. He completed basic, AIT, and Airborne school in early 1975 and got orders for the 2nd Ranger Battalion, the same battalion now commanded by his father-in-law.

John was assigned to 1st Platoon, Bravo Company as a rifleman. After participating in the first ARTEP Combat Certification Evaluation, John attended Ranger School in the winter of 1976. John stayed in the battalion until his enlistment was up and decided to reenlist for Special Forces. Those first two years in the Battalion were some of the hardest physical and mental suffering for all the young Rangers. It taught them the importance of being flexible and adaptable in highly uncertain situations. It also taught them to 'drive on' when the going was so difficult that all they wanted to do was quit. It was two years of steep learning curves and the inculcation of important values that we all would carry with us throughout the rest of our careers and life.

When asked what values were most important to him, he replied, "I think it was honesty, integrity, and loyalty. I think those are a perfect circle. The other values fit into the middle of those three. Honesty, integrity, and loyalty made everything work for us. It built trust and respect. Trust is knowing that when you're given a mission, everybody around you is in the same boat.

"Everybody adopts the same frame of mind. We knew our leaders, especially squad leaders, but also platoon leaders, company commander, and first sergeant, they all had your best interests in mind. You knew no matter what, you were coming back, and if you were injured, somebody's going to take care of you. So, we were able to give it our all, all the time."

After attending the Special Forces Qualification Course in the winter of 1977, John was assigned to the 7th Special Forces Group staff as the training and ammunition NCO. At first, he was disappointed. He wanted to be on an operational detachment, known as an A-Team. In retrospect, John said of his first SF assignment, "I learned a lot about the SF planning process. I could see how good teams planned well and how lazy or incompetent teams didn't. I assisted teams to improve their planning process, to think ahead about what type of support they needed, how much ammo, ranges, training areas, etc."

John reenlisted for 1st Battalion, 10th Special Forces Group in Bad Tolz, Germany because his father-in-law, Colonel Baker was stationed there after leading the 2nd Ranger Battalion. It got them closer to his wife's parents, but more importantly, John knew the way Bo Baker trained us back in the Rangers and that he was a solid leader, which brings us back to the beginning of John's journey and one of the most difficult personal times in his life where his adaptability and resilience would be tested.

After the funeral, John returned to Bad Tolz and got back into the swing of things. Over the next several months though, life threw more unexpected and difficult personal situations at John, resulting in his divorce and his family returning to the States early while he remained in Germany. It left him rudderless. He started losing hope and drinking heavily. As John tells it, "I was just there and everything kind of collapsed on me all at one time. I just wanted to be alone to do some thinking.

"I moved into the team room, we had a small room above the team room upstairs, and I made a little pad for myself. I started drinking. After a couple weeks, two of my teammates came up, and said, 'If you want us as friends, you better pull your head out of your ass.' I dumped the booze down the drain right then, moved into the bachelor quarters, and started all over again. That was somebody helping me, a really good teammate and friend. I decided I needed to get back to see my kids and be nearer to them. I needed a plan. The question was, how was I going to do this? I was only in my first year of a three-year tour."

John got back on his horse and focused on the mission while he tried to figure out how he was going to get back to Ft. Bragg. A few months later, John got his answer. The Army's premiere Counter Terrorism Special Mission Unit, 1st Special Forces Detachment—Delta had sent a recruiting team looking for volunteers. The unit was stationed at Ft. Bragg, near Fayetteville, NC where his kids now lived with their mother.

John interviewed, took and passed the psychological tests and a physical fitness test. He was invited to attend the next selection in the winter of 1981. He attended and was selected for the unit. He returned to Germany awaiting orders over the next month, then processed out of 10th Group and returned to Ft. Bragg.

His world was looking bright again. He had renewed energy and excitement. He was close to his girls and was going to be part of the most elite unit in the Army. The challenge was not over. He attended and successfully completed the arduous six-month operator training course and was assigned to an operational squadron. The unit was filling up with many Rangers from both battalions, including NCOs and officers from Bravo Company. John was in good company with men he trusted and admired.

John spoke of his time at the unit, "After I completed all the training, it was yet another turning point because real things were happening in the world. There were lots of bad guys out there. Middle East terrorist groups were hijacking everything in the early 80s. Terrorist insurgent groups were sprouting like weeds in Central America and the Philippines. Our job was to rescue American citizens. We trained for any scenario imaginable, buses, planes, trains, and all types of multi-story buildings. We trained every day and trained hard until we could do all the mission sets in our sleep, just like we did in the Rangers. I finally felt like, okay, now I'm ready to go to combat and pay this forward.

"We did do quite a few real-world things. There was a lot of Ranger mentality in the unit. I don't mean in the negative Rah-Rah way. I mean the Ranger mentality to accomplish the mission. It was all focused on the mission. Let's work smarter, not harder, but let's work at it. That's how it was."

After five years of running around the world chasing terrorists, participating in real rescue missions and being gone for 200+ days a year, John decided to get out of the Army to have more time with his growing daughters. He took a job as a Department of Defense Civilian at the Special Warfare Center teaching military free fall (MFF) in 1986. The equipment was still dated from the Vietnam Era and John became an integral part of modernizing the equipment and the tactics, techniques, and procedures of military freefall infiltration.

Eventually, John took over the research and development shop for the school and spearheaded the testing and evaluation of equipment, including developing the requirements for new parachutes, specialized equipment, and publishing updated field manuals. Faced with an unresponsive bureaucracy that was more focused on the Commander's pet projects in lieu of the mission, John resigned in 1993 and went back to the fight as a defense contractor, retiring in 2010.

John moved to Asheville, NC with his wife Libby. Today he leads a quiet life of giving back to his community, or as he likes to say, paying it forward. He is a sterling example of resiliency and adaptation. He continues to live his life within his perfect circle of values: honesty, integrity, and loyalty. He teaches others how to overcome challenges through thoughtful adaptation, common sense, and keeping a positive attitude to help them to be resilient when times get tough.

* * *

This brings us to a story of how perseverance, attitude, and faith can lead to a fulfilling life, even when dealing with a lifetime of self-doubt.

It was Beginning Morning Nautical Twilight (BMNT), the military term for pre-dawn. It was always coldest at dawn and this morning was no different. The soaked Rangers were spread out in their assembly area at 100% security in case the enemy had detected them coming while they waited for word to come down to move out. The terrain was still difficult, but not as difficult as it had been the night before with all the deadfall and 'wait a minute' vines during their 4-hour movement to the assembly area. This was it. The final 100-meter push, the rest of the way up the hill to attack a dug-in infantry company. Bravo Company was on the left, Alpha Company on the right and Charlie Company in reserve. The word came down, move out.

2nd Platoon was now first in the order of march for Bravo Company. 3rd Squad was first in the order of march for 2nd Platoon this time. Alpha Team was on point. Specialist 4th Class Bill Waterhouse, the Alpha Team Leader, was told to move out. He was leading the battalion towards their final attack position where they would spread out, on-line for the final deliberate assault on their objective 100 meters away. When Bill reached the release point, he halted and Bravo Company fanned out online, about 10 meters between individuals.

Alpha Company did the same thing on the right. They used the cover of the trees and thick vegetation to conceal their movement from their release point to their final attack position. A separate support element was moving into position on Bravo Company's left for supporting fires from machine guns and mortars. When the attack force was about 30 meters away, the signal was given for the force to rise and rapidly attack. The support element opened fired on the enemy to keep their heads down, providing cover fire for the attacking Rangers.

It was over quickly, the Rangers seized the objective and moved through it, securing the far side, and establishing a defensive line in case of a counterattack. Shortly thereafter, the word came down, ENDEX. End of Exercise. The ARTEP was finally over.

Bill fondly remembers, "I was in B Company, and I was a fire team leader, and I had the best fire team. It all came home on that last movement we were on. We were moving through the woods, and I had nobody in front of me. I'm the fire team leader. I have my fire team in a wedge formation and I'm looking up ahead and I'm looking around. I'm thinking to myself, where is everybody? It was getting light, and I looked behind, I could see our squad's other fire team and our squad leader and further back the rest of our platoon. They were all behind me.

"There was an ARTEP evaluator walking with my team, I'm not sure why he was there, but I asked him, 'Where's the battalion?' He said to me, 'They're all behind you!' As exhausted as I was, it was like day one of the ARTEP. I was standing tall. I was the point man leading a 500-man attack force."

Bill goes on, "A few days later, after we had cleaned up and before we were going on block leave, I remember LTC Baker standing up on the PT stand surrounded by his 500 Rangers. He was telling us how proud he was of us. I remember distinctly him saying, that of the 800 men who had come through the battalion, close to 150 men had come and gone since the battalion was stood up a year before. They left because they couldn't or wouldn't adapt to the high standards, and they didn't have the resilience and resolve to tolerate the suffering we had endured over the past year. We were the last ones standing. I was so proud to be a Ranger that day."

Most of those 500 men were just out of high school, somewhere between 17 and 21 years old. All had volunteered for the Rangers. None of them had been in combat nor ever experienced the kind of physical and mental suffering they had just been through over the past year. Many were high school athletes, but no amount of athletic summer camps and daily high school training regimens could compare with the deprivations these men had lived through for the past year. Those standing there that day had learned resiliency the hard way. The Ranger way. They were the ones who learned to adapt and found ways of learning to endure the grueling physical and mental challenges most of these young men had never been exposed to in their lives.

The ones who stood there that day had learned to accept a single truth about being a Ranger: 'If you don't mind, it don't matter!' Or as enlisted men are apt to do since the creation of the Army, they made up an acronym for it. In this case it was FIDO. 'F*** it! Drive on.' The ultimate tribute to adaptation and resilience.

William Carson Waterhouse Jr. comes from a military family that goes back to the Revolutionary War on his father's side. His dad was in the Army Air Corps in World War II, and in 1946, made the transition to the U.S. Air Force. Bill's childhood was not easy. A child of alcoholics who turned their own self-disgust onto their little boy, who grew up being told he was worthless. Bill managed the hurt by turning to fantasy and playing Army with his friends. When Bill was thirteen, an album by a Special Forces combat veteran, Staff Sergeant Barry Sadler, entitled "The Ballad of the Green Berets," made the charts. Its chorus line inspired Bill to join the Army.

Silver Wings upon their chest,
These are men, America's best.
100 men will test today,
But only three, will win the Green Beret.

Bill decided right then, at 13 years old, to commit to earning those 'Silver Wings Upon His Chest.' Motivated in part by wanting to escape an abusive environment and hoping to find a family that would accept him. When Bill graduated from high school he went down to the Army recruiter. When the recruiter asked him if he was sure about Airborne School, Bill had second thoughts. He lost confidence and feared he couldn't do it, so he opted to join the Army as a wheeled and tracked vehicle mechanic. His scars from years of mental abuse ran deep and undermined his self-worth and confidence. But time, desire, and courage would make those scars into resilient shields and Bill would gain confidence over time.

After he finished mechanic school, he was assigned to the 3rd Battalion 5th Cavalry (CAV) Regiment (3rd/5th) at Ft. Lewis, WA. After three weeks at the unit, he knew he had made a big mistake. In 1973 there were no Ranger Battalions, only independent Ranger Companies of the 75th Rangers. These reconnaissance companies were a relic of the Vietnam war. They were a Corps level strategic reconnaissance asset. At Ft. Lewis it was Company B, 75th Rangers. B Company was attached to 3rd/5th CAV for their logistical support. Bill learned about the Rangers and developed a plan.

In Bill's words, "I saw my opportunity. I went to a barber and got a Ranger haircut, a high and tight. I had a set of fatigues triple starched and bought a new pair of jump boots and polished them until I could see the reflection of my face. I jumped on a post shuttle bus and went to the company headquarters of B Co. I walked into the orderly room and a young second lieutenant asked me, 'What can I do for you, Private?' And I told him I wanted to join his outfit. That was it. He immediately took me out without another word and put me through a PT test. After I passed, he says to me, 'Private, we'll have orders for you in two weeks.' Two weeks later, I was assigned to their motor pool, to service the few jeeps and trucks they had. I was a happy man."

Shortly thereafter, they sent Bill to Airborne School, thus completing the commitment he had made to himself as a thirteen-year-old to get 'silver wings upon his chest.' He adapted and overcame his fears and had the resolve to follow through. Bill was with B Company for about a year, but because there wasn't much vehicle repair or maintenance to be done, he tagged along with the recon teams in the field. He was surprised that the teams liked him and treated him as one of their own.

When the recon company was disbanded in 1975, he learned that the time he spent with the teams in the field over the past year earned him a secondary Military Occupational Skill (MOS) as an infantryman. His new MOS allowed him to volunteer for the 2nd Ranger Battalion, since the Battalion was a light infantry unit and had only two jeeps on their Table of Organization and Equipment (TO&E). He was accepted and arrived in Bravo Company in January 1975 as an E4, Airborne qualified, infantryman. He was assigned to 3rd Squad, 2nd Platoon, Bravo Company as a team leader.

The dark lack of self-confidence began to creep back and erode his belief in himself. He admitted he was very nervous since he had no formal infantry training other than basic training. As Bill described it, "I thank God today that the wisdom of the leadership said, 'We're going to start this battalion off with a mini 11B (Infantry) school.' Literally from the beginning. This is an M16. This is its cyclic rate of fire. For the next eight hours we are going to disassemble and reassemble the rifle and clean it. You're going to do it blindfolded, in the dark, when it's raining, snowing, hailing, in 100-degree weather, whatever, until you can do it in under three minutes. We did this for everything, gas masks, grenades, marksmanship, and 100 other, necessary 11B skills we had to master over the next several months. I didn't know 'fire' from 'forward march,' so I just kept my mouth shut, my ears open, and learned."

Those many months leading up to the December ARTEP, everyone started the journey at the same beginning. Squad leaders and experienced team leaders taught the new privates and inexperienced E3s and E4s. The training was focused on learning and demonstrating the newly acquired individual and squad level tasks they had to master. This is how Bill ended up leading the entire Battalion on its final mission during the ARTEP. The best part of the story, Bill wasn't even a 'tabbed' Ranger, yet.

Bill was selected to attend Ranger School in January 1976. Unfortunately, he ended up getting very ill and was held over until the April 1976 class. He successfully completed the eight-week and five-day trial, which changed his life forever. Bill describes his graduation from Ranger School, "I grew up in a very difficult situation. There wasn't a lot of 'lifting up' or positivity growing up. I was told I would never amount to anything. They'd say things like, 'You're a rotter. You're this, you're that.' It was all drunken talk that my parents eventually resolved later in life. But the damage was done to the little 13-year-old teenager.

The Army became my family. I found my home in the Battalion. But when I graduated from Ranger school, something changed. When they pinned that black and gold tab on me, I left that demon behind, or so I thought. I realized my parents were wrong and I had some self-worth. But upon returning to the battalion as a sergeant, the self-doubt returned that I somehow still was not good enough. I think it might have been that I scored very low and had barely passed Ranger School. I had to find myself again, that's why I left the battalion and went to Germany."

Bill was assigned to the 1/15 Mechanized Infantry Battalion, 3rd Infantry Division. Because his primary MOS was still a mechanic, he was assigned to the motor pool. It was a culture shock, the antithesis of the Rangers. Total lack of individual and unit discipline, apathetic and incompetent leadership, totally dysfunctional units. After two months, Bill went to his First Sergeant and told him, "I can't do this. This is not where I need to be. I'm an infantry squad leader." He reassigned me to an infantry platoon to be a squad leader."

After a few months, Bill assessed the situation. "We weren't doing any training, we were being "detailed out" for mess duty, post support, picking up trash, loading trucks, you name it, anything but training. It was so bad, I had to learn to be bold in a respectful way. My Platoon Sergeant wouldn't listen, didn't care, neither did the First Sergeant. They were both 'Retired on Active Duty' (ROAD), waiting to hit their 20-year mark.

"So, I went to my Company Commander using the open-door policy. I was that desperate. I said to him, 'Sir, you're expecting us to move to the Fulda Gap and defend Western Europe from the big red horde and we're not ready to go to war, Captain. Give me my squad for two months and I'll have them ready.' Well, my First Sergeant and Platoon Sergeant had it out for me after that. I was making them look bad."

After two months, his squad was assembled with all of the other squads from the entire battalion at a rifle range to conduct a live fire attack exercise using proper fire and maneuver, cover and concealment, across 300 yards. They were firing live ammunition against plastic 'pop-up' targets, representing enemy soldiers. The competition required the squads to maneuver to the end of the range and secure the pretend objective. This was a timed event from the time the squad commenced until they completed the live fire exercise. The squads were evaluated for their proper use of small unit tactics and marksmanship accuracy. Only two squads completed the course successfully, his and one other. He and his squad were ecstatic. Their morale skyrocketed. They'd never been prouder of anything they'd done since joining the Army. However, Bill did not get the same reception from his First Sergeant or Platoon Sergeant.

Bill recalls the reaction from his platoon sergeant, "My platoon sergeant called me to his office and said, 'I can't fire you, Sergeant, because you're not incompetent, but I'll do whatever I have to, including lying, to screw you.' At that point, my platoon leader walked in, and I said, "Sir, there's nothing more I can do here. With your permission I want to talk to the Battalion Command Sergeant Major (CSM)."

CSMs know what's going on in their battalion. I met with him and explained the situation. As I stood there, he called the First Sergeant and told him to make me the Platoon Sergeant immediately or he would move me. I was an E5, a junior sergeant. My Platoon Sergeant was an E7. The CSM wanted me to replace the E7."

In the end, the CSM pulled Bill up to Headquarters where he worked directly for him coordinating the battalion's movements and supplies for the next 18 months. Bill learned a lot about staff work, but what would transform his life forever was learning about the Army's Warrant Officer aviation program. It would require him to take the Flight Aptitude Selection Test (FAST) and flight physical. If he passed them, he would attend the Warrant Officer Training Program and then attend Rotary Wing, Aviator School, to become a helicopter pilot. Bill told us, "I investigated it. I met the basic Army aptitude requirements, so I went and got a private pilot's manual of basic aeronautics and studied the heck out of it and took the test. I scored very high. I passed the flight physical with no problem, applied, and got orders for flight school."

Excelling in the toxic work environment in Germany, Bill's confidence and self-worth were on repair, his resilience growing. Doing so well on the flight test really gave him a shot in the arm. He attended Warrant Officer training and did very well, same with flight training, and he graduated as a UH-1H "Huey" helicopter pilot. He was assigned to Ft. Polk, LA with the 5th Infantry Division aviation detachment.

As Bill tells it, "I had graduated from Ranger school, so I knew I could do anything. After I'd been there awhile, I kept looking at the AH-1, Attack Helicopters, known as Cobras, sitting on the tarmac. I really wanted to do that. I asked my company commander if I could go to Cobra school. He says, 'Well, you can try, but nobody has ever been successful.' I said, 'Well, I like the challenge. Tell me I can't do something, I'll do it.'"

Bill went through the process and was accepted, got his orders, and upon successfully completing the Cobra transition course was assigned to the 4th/12th Calvary Squadron at Ft. Polk. He graduated from attack helicopter school and was qualified on four models of fully modernized Cobras. He loved it and excelled as a pilot. He was selected by a flight instructor evaluator from the World-Wide Standards at Ft. Rucker for a no notice pilot evaluation in a UH-1H Huey. He passed the check ride with high marks but once again, life threw a wrench into Bill's plans six months later.

It was 1986 and newly passed legislation known as the Graham Rudman Act required the federal government to reduce its $165 Billion deficit and balance the annual budget by 1991. To do this, a mechanism for automatic spending cuts called sequestration was triggered to reach deficit targets.

Bill, and many other Army officers and senior NCOs, got caught up in the cuts either due to lack of civilian education (in Bill's case) or they had their 20 years in. Without warning, he was involuntarily discharged from the Army, with a $20,000 separation pay, an Army Commendation Medal to show the Army's appreciation of his 13-years of active-duty service. However, prior planning prevents poor performance. Bill had a plan, so he returned home to California with his wife and baby boy.

Bill applied to become a California Highway Patrolman (CHP) with the goal of serving in their aviation unit. He attended CHP Academy and spent two years as a patrol officer. He applied to, and was accepted by, their aviation division. He flew for a year and decided to return to the ground patrol division. It's a Ranger thing. The Patrol Division was a lot more interesting. He thrived as a highway patrolman for the next 10 years. He was recognized for his excellent drug interdiction work, which earned him a spot on a multiagency drug interdiction task force.

He was brought onto the state board to evaluate and coordinate interdiction training for patrol officers and taught multiple different local and federal law enforcement agencies on strategies and techniques to identify drug traffickers in transit. Bill had decided to return to the aviation division and was accepted. Unfortunately, before he could join the division, he was injured on the job pursuing a fleeing suspect and forced to medically retire after 14 years with the California Highway Patrol in 2001.

Bill's resilience and ability to adapt helped him find his new calling as an Emergency Room/ICU technician in his small, northern California town. He worked there for the next 20 years, retiring in 2021. Today, Bill is the spiritual leader in his small church. His new mission is assisting his congregation to deal with life, faith, and service to others. Bill is the living embodiment of adaptation and resilience. Having lived his whole life fighting depression from his early childhood experiences, his inspirational story is one of resilient adaptation through faith and work.

The stories of John Snape, John Funderburk, and Bill Waterhouse demonstrate that with resolve, drive, integrity, courage, and honor, one can overcome just about anything in life if you are willing to accept your circumstances (resilience) and then find a new path forward (adaptation). In the age-old adage of, 'when one door closes, another opens,' these Ranger stories represent their utmost belief in themselves and their core values the Army helped reinforce to provide them the strength to overcome and excel, even after some of their darkest days. It is the can-do spirit. You might have to take a knee for a while, but then it is back on the horse. "FIDO."

7. TRAINING TO HIGH STANDARDS

For training to be effective, it must be measured against established standards. If there are no standards, then you are wasting time and resources and not contributing to mission readiness.
—CPT Lawson Magruder

From upper left: Ron Buffkin, John Brasher, Dave Hill, Robert Demoisey

SHORTLY AFTER BRAVO COMPANY WAS FORMED, LTC BO BAKER, OUR BATtalion commander, was told that our First Sergeant, Bill Block, was an expert in hand-to-hand combat and had previously sponsored the demonstration team at the Ranger Department. Bill was tasked to form a demonstration team to have it ready for our Nation's Bicentennial celebration at Fort Lewis on July 4th, 1975. Bill trained the team after duty hours and had it ready for the celebration. His carefully scripted and choreographed demonstration stole the show and was used many times when foreign dignitaries visited Fort Lewis. After Bill retired in late 1975, leadership of the team was turned over to Sgt. Frank Rekasis. The subsequent leader was Sgt Fred Kleibacker who led the team until 1978 when he was reassigned to Ft. Bragg. Fred needed a strong young leader to be his substitute. Fred selected a member of the team, Ron Buffkin, to take his place. Little did Fred know how the selection would change Ron forever...

Ron Buffkin: "SGT Fred Kleibacker was the leader of the hand-to-hand combat demonstration team. This was after Sgt Rekasis, who had led the team, departed. The hand-to-hand demonstration team featured a precise script with exact timing cues. Fred picked me to replace him as the narrator, with one catch. I had to memorize the script and prove to him I could narrate the demonstration. I was already a member of the team and knew all the moves but had never considered leading it. I was a poor student in high school, but I wanted the hand-to-hand leader job badly.

"So, I studied and memorized the script and Fred picked me to replace him. That one hand-to-hand demo script 'homework' assignment from Fred triggered a study skill I possess to this day. I can memorize easily if I want to. Being a helicopter pilot requires lots of memorization to learn and ingrain emergency procedures, flight characteristics, and aviation knowledge. I also used this in pursuing and receiving three graduate degrees later in life. It taught me that if I want something, I have to work for it."

Training was the lifeblood of 2nd Rangers and Bravo Company during the first year. Everything we did was focused on the words of General Abrams in his original charter for the Ranger battalions: "The battalion is to be an elite, light, and the most proficient infantry battalion in the world. A battalion that can do things with its hands and weapons better than anyone."

We could not waste our time on tasks that did not improve our individual and collective skills. We decided we needed to train to a standard and not to a time. If a task did not have a standard, along with the conditions under which it would be accomplished, we developed and documented the standard. Soon the entire army would adopt what was soon referred to as performance-oriented training.

After every mission, we would sit down and provide feedback at every level to improve performance. These sessions would sometimes be quite brutal, but always delivered in a respectful tone. Later, the Army would call these "after action reviews."

That first year was extremely exciting because we were provided the time and resources to focus on our goal of combat readiness. We took a progressive approach to the training calendar, spending equal number of weeks on individual, squad, platoon, and company level training before we progressed to battalion synchronized operations.

There were no distractions during that first year other than the continuous arrival of new soldiers who needed to be integrated into the team and individually trained by their NCO leaders. This took supreme effort with after duty hours and weekends spent with new soldiers. Four of those young Rangers were Ron Buffkin, John Brasher, Dave Hill, and Robert Demoisey. Three arrived raw and inexperienced, but they were disciplined and eager to learn. Of note, after they left Bravo Company, they went on to become extraordinary trainers over the next five decades. **Here are their stories:**

Ron Buffkin knew at an incredibly early age that he wanted to be a soldier. His dad served in the Korean War and was immensely proud of his service and the positive impact that the Army had on him. Ron watched every episode of the series "Combat" and remembers a coloring book he cherished that was focused on Rangers. He knew he would eventually go to college, but straight out of high school he enlisted into the Army. His initial enlistment was for three years in the Army Security Agency. A recruiter told him that it would be extremely exciting work and that he would be able to jump out of airplanes.

But shortly after he started his advanced schooling at Fort Devens, his position was abolished, and he needed to find a home. An article in *Soldier* magazine that talked about activation of the 2nd Ranger Battalion caught his attention and he soon was headed to Fort Lewis in hopes of joining the battalion. At the replacement detachment was a senior NCO from the battalion who told the new soldiers about the unit. After the presentation, Ron immediately signed up. Little did he know that that would start his 33-year journey as a soldier.

Ron spent three exciting years of growth in Bravo Company. He experienced training in a variety of environments: subarctic, jungle, desert, and water. He learned how to train to standard under the most challenging conditions. He learned the importance of being personally competent to train others. The NCO credo of; "Be, know, do" was demonstrated every day by his superiors from his company commander all the way down to the most junior NCO. Here are just a few of the key leadership lessons that Ron took away from his formative years:

On developing others: "It's the leadership value of developing your subordinates. And it's something that every leadership manual, doctrine, leadership book talks about developing your people. It takes a leader to look deep in his formation and reach out and pull up and develop subordinates. That requires a deep personal insight into the strengths and weaknesses of your subordinates, to reach out and give them something that's harder than they would do on their own."

On relationships: "Like I tell people, the first platoon I was assigned to permanent party in the Army (in the 2nd Ranger Battalion) was a platoon of all pretty much single men under the age of twenty-one. I never had so many friends in my life as I did in the Ranger Company. In the rest of my army career, even in aviation, even special operations, even the units I helped form and create, I kept trying to recreate that brotherhood. And it was not a matter of us being the same age, the same ethnicity. It was a matter of us being steeped in the same discipline, training, environment, and having to endure suffering with a group of people, those who become your blood brothers."

On training to standard: "And I think those experiences, they can't be formed in an academic environment. They have to be formed under conditions of hardship, like in the Rangers, when you have those peers who are expected to live up to a standard in all things, and you endure that with them together, which at that time, the rest of the Army wasn't doing anything like that at all. But from the moment we woke our eyes up until we put our pillows on our bunk beds with our OD blankets and went to sleep, we were expected to live up to a standard. All those Rangers lived up to that standard."

On caring for others: Ron remembers vividly the day he arrived in the company and how he was warmly greeted not only by his platoon sergeant, Roy Smith, and squad leader, Fred Kleibacker, but his company commander, Captain Magruder, who shook his hand. This care was displayed over the next three years through quality training in a feedback rich environment.

After Ron's enlistment ended, he decided to stay in the Army and earn a commission, then follow his dream to fly helicopters. After commissioning from Officer Candidate School and a couple of years earning his undergraduate degree, he went to flight school followed by an assignment in Germany flying the OH-58 Scout helicopter. It was there that he learned that many aviation units were focused only on the technical side of the business and not tactical leadership and development of others.

Somewhat disillusioned, he volunteered for and was accepted into the Army's new special operations aviation unit, the 160th Special Operations Aviation Regiment at Fort Campbell, Kentucky, after completing the Infantry Company Commanders Course he volunteered for. It was here that Ron would serve for three years with Army Aviation's most skilled aviators as they supported special operations units from all services. Their helicopters were the most advanced in the world, requiring the most competent leaders and pilots to fly them under the most dangerous conditions.

The training, planning, preparation, and execution by the 160th reminded Ron of what he had learned as a young Ranger.

The 160th assignment reenergized Ron and propelled his career. He commanded aviation units worldwide at every level from company to battalion, culminating his career as a full colonel in command of a brigade. At every stop, he would harken back to his time as a young, enlisted Ranger. He would lead the way with his technical competence, tactical moxie, and his compassionate leadership. He would instill into his units many lessons he learned as a young soldier and NCO. The most important revolved around "duty."

His words capture its true meaning: "In Bravo Company, we were imbued with the sense of executing your duty, whether that's by yourself, for your fire team, for your company, for your battalion, for your army, and for your nation. And I would say that's one thing that I carried the rest of my time throughout the Army, and that's what I looked for in my subordinates and commanders. Did they have a sense of duty?

"That sense of duty is what we had in the battalion. We also developed leaders. If you think about the way the Rangers were structured, you went to Ranger school as a 19-year-old PFC, and Ranger school is a leadership school. And in that environment, you were forced to step up as a leader, even though you're a private. That's another value I looked for, encouraged, and sought to develop throughout my Army career. And there's a couple of little sayings. Napoleon said, 'There's a field marshal's baton in every private's knapsack' and in aviation units, 'Your value to the unit is not determined by your proximity to the cockpit.' The young private avionics technician is just as important from a leadership perspective as the standardization instructor. Everyone in the outfit is important to mission accomplishment."

* * *

John Brasher and Dave Hill followed similar paths in their journeys over the past 50 years. Both went on to be commissioned as Infantry lieutenants. Both spent their commissioned years as trainers. Both retired as officers and both continued to serve as trainers in their civilian jobs. Both clearly reflected General Abrams' guidance that: "The values and standards established in the Ranger Battalions would be spread to the rest of the Army with the assignments of its veterans to units throughout the force."

John Brasher's father also served in World War II and was his hero. After high school, John was working as a carpenter and making good money but was not managing it well. He had attended college for 18 months but did not do well. He decided to enlist because he was offered a $2,500 enlistment bonus. (He was surprised when he only got a check for $2,000 because 500 was taken out for taxes!) He enlisted under the Ranger Option for three years and flourished in 3d Platoon under the leadership of Lt Eldon Bargewell, Platoon Sergeant Roy Smith, and squad leader Kim Maxin. He served two years in a Ranger squad, advancing from rifleman to team leader to squad leader. He then wanted to experience Europe, so he moved to Italy to the 509th Airborne Battalion.

Now a Staff Sergeant, he reenlisted for Fort Benning where he served for two years passing on his experience as a Ranger instructor/trainer. While an instructor, he attended college at night and earned his degree. He was then positively influenced by his Captain superior to apply for Officer Candidate School (OCS). He was quickly accepted for attendance.

John flourished in OCS and was the Honor Graduate. He said the attention to detail and exacting standards from Bravo Company made the demanding course easy for him to navigate. Commissioned in the fall of 1981, John served for 14 more years as an officer in tactical and training units. Unfortunately, he and many others in his year group got caught up in the Reduction in Force or RIF that occurred after the Army was downsized after the end of the Cold War. He retired in 1995 with 20 years of proud service. John's experience as a trainer was put to skillful use for 26 years after he retired from the military. He became one of the preeminent experts on urban warfare, helping in the development of a state-of-the-art

Military Operations in Urban Terrain or MOUT site at Fort Benning. His civilian defense industry company developed training manuals for NATO military organizations. This doctrine is serving an incredible purpose today as we watch the brutal wars in Ukraine and Gaza.

When John reflected on his time in Bravo Company, the following story spoke to the importance of feedback and personal encouragement in which John has excelled all his life: "The first morning I was in the Battalion, we did PT and went on a run, and I thought I was in pretty good shape. I had just come out of basic, AIT, and jump school with no leave in between. I mean, we flew straight from Benning up to Fort Lewis and did our in-processing. The only reason I didn't fall out of the run is because I had absolutely no idea where I was. And when we finally came back around the Battalion area, I did fall out. I don't know how much farther the platoon ran, but I fell out.

"And the first Ranger that I met, who had no clue as to who I was, was Major Bill Powell, the Battalion Executive Officer. He stopped me, talked to me, gave me a little counseling, and gave me some encouragement. Even when we got finished, I didn't know who he was, but that has always stuck with me as the XO took the time to talk to me, a mere private, and give me some encouragement and wish me well."

* * *

Dave Hill's Ranger experience provided him with the standards to work toward for the past 50 years and will guide him for the rest of his life. "I graduated from high school in 1972 at seventeen, had no direction ... rudderless. I was good at staying out of trouble, had two jobs, a car, a girlfriend, and some money, but wasn't satisfied." He thought he would go to college eventually but was not prepared. "I didn't have the means, the discipline, or a goal to strive for." His dad was an enlisted soldier in the Army, his grandfather served in the Army during WW I in the Ohio National Guard, and his uncle as an Army aviator during WW II; all of them encouraged him to go to college. Dave's father learned electronics in the Army to operate and maintain the nascent counter artillery radar in the 1950's and built that experience into a successful career with IBM. His grandfather became an engineer and told him not go into the Marines (for reasons never explained) and his uncle fired him from a no skill job in his construction company to help nudge him into a better future.

He tried college part time and utterly failed. "I needed an adventure, an eventual path to college, and to do something important." He was inspired by a former Sailor to go into the Navy but, long story short, Dave enlisted for the Ranger option in the newly activated 2nd Ranger Battalion at Fort Lewis, WA ... if he could meet the standard, and that made all the difference.

Basic training, AIT, Airborne School, a good GT score and an aptitude for computers that he learned from his dad, landed him in Bravo Company, Weapons Platoon as a mortarman. He was an exceptional mortar man: smart, physically fit, and intensely devoted to the company. In Bravo Company he learned that standards are essential: to be a soldier in a high performing unit, for providing accurate and timely indirect fires, to conduct airborne operations, for executing a road march ... for anything to get done safely, effectively, and at full speed, you need standards. Rising through the ranks to serve as a non-commissioned officer in Bravo Company set his rudder forever.

After his three-year enlistment ended, he went home, enrolled at the University of Maryland and was accepted to the Army Reserve with 11th Special Forces Group because of his Ranger experience. Being a member of the only A Team co-located with HQ also led him to become a member of the 97th ARCOM parachute demonstration team where high standards had a particularly important role. An Army buddy encouraged him to join ROTC to become an officer and at ROTC summer camp in Fort Bragg, NC his passion for active duty was reignited. He wanted to be a platoon leader and to achieve the standards the officers in the Rangers had set for him. Dave was commissioned as a Lieutenant of Infantry in 1982.

He would spend the next 12 years on active duty as a trainer in many roles: mechanized infantry platoon leader in Germany; light infantry company commander in Lawson Magruder's brigade in Hawaii; as an observer controller at the Joint Readiness Training Center (JRTC) in the early days at Fort Chaffee, Arkansas, and then as a doctrine and tactics instructor at the field artillery school in Fort Sill, Oklahoma. Like John, Dave was in an overstrength year group and, with enough time to retire from active duty, he turned in his uniform for civvies and began working as a DA civilian at the newly relocated JRTC in Fort Polk, LA.

For the past 30 years, Dave has built on what he learned in the Ranger Battalion. That experience enabled him to excel as a simulations and instrumentation expert at JRTC with a passion for training and help JRTC become one of the crown jewels of Army training before the first Gulf War and into the post-9/11 era.

Thanks to the expertise gained with the Army, he became a plank owner in the Joint National Training Capability (JNTC) at US Joint Forces Command in Suffolk, Virginia and was subsequently honored as a Distinguished Member of JRTC Operations Group.

Dave's years in Suffolk advocating for joint training and interoperability between the Services and Combatant Commands have been truly rewarding and purpose filled. He credits everything he has achieved to the start he got with the Army and his service in Bravo Company. Now as a member of the Joint Staff, J7 he is sharing his experience with the next generation of joint warfighters to continuously raise the training standards for the entire joint team. Dave is meeting the current challenge to improve joint training and readiness for the joint force of the future by building on the foundational standards he learned as a young Ranger in the Army years ago.

This fuels his 'why': "My why is to make sure that nobody goes into war unprepared, and by training people when they're young, enabling them to make moral, ethical, and correct decisions so that they avoid getting into unjust wars that we can't win. That is what I am doing, and it all got started way back in Bravo Rangers. I decided years ago, I'm not in it for the money. I am in it for the mission. And there's nothing I would rather do than what I'm doing."

* * *

Unlike Ron Buffkin, John Brasher, and Dave Hill, Robert Demoisey arrived at Bravo Company as a senior NCO. He brought a wealth of tactical and technical expertise to bear as a trainer from three tours in Vietnam and seven years of service. He was a former combat engineer who directly supported the infantry in combat. He was not only an expert in demolitions, but also had vast experience in small unit tactics. Simply stated, when Robert spoke, others listened. He wanted young Rangers and his officer platoon leaders to learn and grow under his tutelage. Unfortunately, a couple of his platoon leaders failed to listen to him and did not last in the unit.

Robert was the definition of the competent NCO: combat experienced, highly skilled, self-disciplined, passionate, and positive about training others. He was also a gifted communicator with a very distinctive deep command voice. When he led physical training for the entire company there was no doubt who was in charge, although when he was accompanied by Toby, his Springer Spaniel, we all thought the canine was equally in charge during the long runs!

Robert believed once a young Ranger had been trained to standard, he should be allowed to do his job and then challenged to expand his skill level at the appropriate time. He was all about standards and building trust. Here are his sage words about building trust and teamwork in your organization: "You don't give anybody an assignment, and they go out and intentionally screw it up. Nobody did that. So, you had to trust them. You had to give them room to fail and fail in training. But they learned from it, and that went across the board, whether it was within the platoon or whether it was in the company.

97

"If you gave us an operation order and we passed it down, it was expected to be done. There were no questions. If there was a question about something, then you had the freedom to ask questions.

"I remember one time one platoon was given part of an operation, and I was given another one. And when the platoon sergeant sat down, we agreed that the other platoon did something better than our platoon. And we came back to the commander and said, can we switch jobs? And here's why. And they trusted us for our judgment on it. That was neither one of us stepping down, but doing what was best for the operation. And so, trust was a multiple direction thing. You had to trust each other. You had to trust the squads to accomplish what they did.

"And the way the battalion set it up with the individual training, fire team, squad, platoon, and then married it all together with company operations was a brilliant stroke. So, we grew up together. We grew to trust each other. We also grew to understand there's times for a discussion, and there's times to 'just do it.'"

"There was great latitude for us to operate because of that level of trust and because people understood that we had to be able to execute, and you couldn't look over everybody's shoulder. So that leads to the fact that we respected one another. And then I think the value of honor catches on because we didn't want to let anybody down, right? And everybody made mistakes, but because of the way we were trained, and because of the trust from the leadership when you did make a mistake, we're not going to chew them out in public. We'll bring him in, we'll explain to him what went wrong, make the corrections, and then out he goes. Criticize in private, praise in public, which builds integrity amongst the group.

"You can't hold a grudge. Nobody intentionally went out to screw something up. If they screwed something up in training, it was a learning experience, and you had to mentor them as to what could have been done or maybe should have been done. I know that they did their best, but somewhere there was a glitch. The one thing that's in short supply when you're preparing is time. You just run out of time, but at the same time, you never were done. Preparation. You can hone it and you can hone it again to become more skilled at your individual and team jobs. One person can't do it all."

When Robert looked back on his three years in the battalion, he says here are the primary lessons he learned that he applied in the future: "Take care of the people around you and they will take care of you. No matter how hard or tough you are on your people, they have to know that you love and care for them. Trust in them, give each one the space to learn and grow, set the example and never stop learning. Make yourself a Subject Matter Expert (SME) in as many areas as you are able to, thus making yourself more valuable to the organization as a whole."

Robert spent the next 10 years of his career as a trainer in several different organizations. After completing the Special Forces Qualification course, he was an instructor at the Officer SF Course, and then an A Team Engineer specialist in the 10th Special Forces Group in Germany. His final years on active duty were back at Fort Lewis. He was the First Sergeant for a deep reconnaissance unit formed as part of the Army's experimentation of new doctrine, equipment, and units. His final year he helped form the 1st Special Forces Group as an A Team Sergeant (Master Sergeant).

With his vast combat and training experience, Robert naturally gravitated to the training arena as a civilian. He held a variety of positions. As a contractor and then a police officer in Thurston County, Washington State, he, and another retired senior Ranger NCO trained the SWAT team specialists in the police department. After a stint as an undercover narcotics police officer, he hired on as an operator and trainer in support of the War on Terrorism. He served in several dangerous locations to include Pakistan and Afghanistan.

At each location, he harkened back to his experience as a Ranger platoon sergeant: "I think the ability to work together to get the job done, having trust in each other, identifying weak points and strong points, combined with a lot of patience, because after a time they would rotate military section leaders. I remember there was a guy that was in charge of the base.

He was master chief out of the Navy, and he was a legend in the SEAL community. He was an older guy, and somebody said, we had explosives. And I went to him, and I said, has anybody looked at those lately? And he said, well, no. And he found a key, and I found the locker it was in. And I went back, and I told the men, don't store blasting caps with explosives. This is bad juju.

"A bit later, we got sniper rifles. So, I told him I have experience as a sniper. He goes, 'Damn, Demoisey, what are you, the Swiss army knife of the military?' I said, 'Well, that's the way we were trained. To be more valuable to whatever job we come to, so we haven't scratched the surface yet. There are other things that I might be able to be of assistance on.' And he was a great guy. I remember one time he said, 'Damn. I love you Rangers, you guys always get the job done and do what I asked for, and I don't have to ask you twice.' I took that as a great compliment for all Rangers that he'd encountered."

Ron Buffkin, John Brasher, Dave Hill, and Robert Demoisey are shining examples of "trainers for life." But the same can be said for every other Bravo Ranger we interviewed. They each became expert trainers in their field of expertise, significantly benefiting their community and our Nation.

"The principles of leadership in the military are the same as they are in business, and church, and elsewhere: a. Learn your job b. Train your people and c. Inspect frequently to see that the job is being done properly."
—Admiral Hyman G. Rickover

8. COMPETENCE

There are four key elements to competence:
knowledge, skill, discipline, and attitude.
— General Richard E. Cavazos

From upper left: Bob Williams, Andy Pancho, Cliff Lewis

1977 — FORT LEWIS, WASHINGTON. BOB WILLIAMS AND HIS MACHINE GUN crew had just spent a couple of hours preparing their defensive position. It was complete with overhead cover and range card. They were very proud that it was to Ranger standards. As they cleaned up around the position, who should walk up but Major General Dick Cavazos, the Commanding General of the 9th Infantry Division, and the Bravo Company Commander. In Bob's words: "General Cavazos talked to us, and we explained our left and right limits of fire, our dead space, maximum and minimum ranges of the machine gun, showed him our range card, showed him where our left and right limits were, where the left and right squads were, and talked about our primary and supplementary positions. He was very impressed with our position and complimented us on our hard work and our competence. He then moved on without saying a word.

The next day, the company commander came up to me and he said, "You know, General Cavazos just about had tears in his eyes and said, the Army doesn't deserve soldiers this good."

(Author note: Fort Cavazos, formerly Fort Hood, is named after General Cavazos who earned the second highest award for valor, the Distinguished Service Cross, in the Korean War and the Vietnam War for his courage leading soldiers in the defense. He is the first Hispanic 4-star general in the history of the United States Army).

General Cavazos' definition of competence captures the key ingredients we were trying to instill in every young Ranger and leader in the intensive first year of training in Bravo Company. The leaders brought a variety of knowledge from formal military schools, and experiences from other units, and some from combat in Vietnam. They also had varying levels of skill gained from those experiences.

To bring everyone up to the same level of knowledge and skill, exacting standards were developed for the key tasks and missions the unit would be asked to perform. It took constant practice and repetition during field training under the most demanding conditions to forge technical and tactical competence.

Concurrently, the intangible elements of competence, discipline, and a positive, can-do attitude, were being instilled at every level. Four Rangers have defined competence throughout their adult lives: Bob Williams, Brian Quinlan, Andy Pancho, and CC Lewis. Here are their stories:

Bob Williams was inspired to serve in the military by his father who served in combat in the Korean War. Bob's father joined the Army when he was 17. He joined because he got into some trouble. He was an adventurous young man who selected the airborne option, ended up going to the 11th Airborne Division at Fort Campbell, Kentucky. Once the Korean

War started, he wanted to be a part of the action and joined the 7th Regiment of the 3rd Infantry Division which coincidentally was the same unit in which author Lawson Magruder's dad served.

As related by Bob: "In one battle, his platoon got overrun by Chinese, and everyone was killed except for my father and two other soldiers. The only way they survived was they got trapped in a cave bunker, and when the Chinese would throw grenades in and try to rush them, they would toss the grenades back. They would use small arms fire until they ran out. And then it was hand to hand combat. They managed to hold back a battalion of Chinese until the rest of his regiment could conduct a counterattack." Bob said after the war his father left the Army and worked for the railroad to help raise six children.

The exploits of Bob's dad were captured in memorabilia, pictures, and articles that excited him as a youngster. After high school he was working in a factory in Manistee, in the northwestern lower Peninsula of Michigan, and was trying to save up money to go to college, but he admitted he was having too much fun with his friends who didn't want to go to college. He soon realized that factory work wasn't going to do it for him, so he went down to the local recruiting station to inquire about the Army.

He was very fortunate to have a recruiter who was a proud Vietnam veteran Special Forces NCO. His recruiter was incredibly careful about helping Bob decide on where to serve. They both decided that the 2nd Ranger Battalion was a perfect fit for his sense of adventure and love of the outdoors.

Bob spent three intensely developmental years in 3d Platoon of Bravo Company. Arriving as a PFC fresh out of airborne school, he took every opportunity to learn from outstanding leaders and to gain knowledge from formal schools and training exercises and deployments. He graduated from Ranger School, Jumpmaster School, the Scout Swimmer Course, PNOC (Primary Non-Commissioned Officers Course, BNOC (Basic Non-Commissioned Officers Course), and a technical mountain climbing school. He rose in rank from an ammunition bearer in a machine gun team to a squad leader leading nine other Rangers.

He had two wonderful role models in his platoon leader, Lieutenant Eldon Bargewell, and Platoon Sergeant Roy Smith. Much has been written about Eldon in other chapters, but Roy Smith must be mentioned about his positive impact on the young Rangers like Bob. Here's what Bob had to say about Smith: "He was my very first platoon sergeant. He was quite humble and soft spoken, but he was a big guy. I really got a chance to know him one to one during our time together. When we had down time in the field and we had time to just talk, he was very approachable. We talked about the military, but we also talked about family. And as a young Ranger, to have an NCO like that really sit down and just give me the time of day and talk about his family and kids was pretty special to me."

Lawson Magruder reiterated Bob's comments about Roy Smith: "He always impressed me with his humility. He was soft spoken, but as the company commander, I saw him behind the scenes doing what a platoon sergeant is supposed to do while supporting a strong leader like Eldon Bargewell. The third platoon did exceptionally well, but it wasn't all Eldon Bargewell. It was also a platoon sergeant who knew his business and kept squad leaders and team leaders on track. I greatly admired Roy Smith and his family."

Bob's experience and schooling in the battalion led to him becoming a truly competent soldier and leader. An important ingredient of competence is self-discipline, about which Bob shares: "Probably the biggest thing I got out of the Ranger battalion was self-discipline. Tied directly to that is loyalty to my teammates, not wanting to let anybody down. When I had situations where I felt weak or that I couldn't measure up, I used the peer pressure that I held within me, knowing that I'd let people down if I didn't do the job, to motivate me. And part of that was competition, admittedly. This loyalty forms and forges something even deeper, I guess a friendship that encompasses a value of respecting somebody, trusting them, and, wanting to do good for them, knowing that they're going to do good for you."

When Bob reflects on his three years in the battalion he states: "I didn't know what I didn't know when I got there. There were things I didn't understand because I just didn't have a clue. But it was both a melting pot and a forge because it took young men like me that were kind of coming into our own and shaped us into men that we wanted to be. And the best way to do that is to be surrounded by men that you want to be like."

Bob got out of the Army after his enlistment ended and went back to Michigan to college. After a short stint as a big game guide in Idaho, he enrolled at West Shore Community college, focused on a degree in marketing management. One day he was headed to class when a chance meeting with a fellow whose car was broken down on the side of the road led him into a 26-year career in the Army National Guard.

Bob related: "I pulled over to stop and help him, and he was looking at me, and we were just kind of chatting while we were getting his tire fixed, and he asked me if I was in the military. I said, no, I just got out. I'm going to college now. And he said, "You know, we got a National Guard unit here in town."

I said I didn't know. And he said, "I'm getting ready to go to Officer Candidate School (OCS). Why don't you join that unit and go with me to OCS? I looked at him, and before I could even answer him, he said, "Look, it'll help you with your management experience for college." He just kept talking. By the time he was done, I actually believed him. So, the next day, I went down to the local Army recruiter and told him I wanted to join the local guard armor unit. And so that's how I started my national guard career."

Bob flourished in the National Guard. After commissioning as a lieutenant, he served in many leadership positions primarily focused on the training arena. He had several key positions in the Michigan Army National Guard: Tactical Officer and Tactics Instructor in the Michigan Military Academy, platoon leader, operations officer, and executive officer in a Long-Range Surveillance Company, Instructor, S3 operations officer, Ranger Challenge Coach, and Commandant of Cadets at the University of Detroit Army ROTC program. As a company grade officer, he focused young soldiers, cadets, and leaders on high standards and becoming truly competent in their appointed position. Simultaneously, he was earning an undergraduate degree from Michigan State University and later a Master's Degree from the University of Arkansas.

As a field grade Active Guard and Reserve Officer (AGR), Bob's competencies in the training and education arenas were maximized in several impactful assignments: In Germany he was the Overseas Deployment Training manager for the U.S. Army; he was an instructor and chief of the training center at the Army National Guard Professional Education Center at Camp Robinson, Arkansas; as a full colonel he was the Senior Army National Guard advisor to the Seventh Army Training Command in Germany and later the Senior Army National Guard Advisor to the Commanding General of U.S. Army Europe.

One of the highlights of Bob's senior leader experience in Germany were his contributions for the development of the concept for and later the Army's establishment of a Joint Multinational Readiness Center. It is the only combat training center outside the United States. It is one of the "crown jewels" and centerpieces for unit training in the Army.

When Bob speaks about the behaviors inculcated in him as a young Ranger and how he applied them later in his career, he states: "I learned how to become a good observer. I can observe people and their behaviors, situations, and come up with a reasonable assessment. I can recall my lessons learned through experiential learning, specifically theory and experience, and can apply them to most situations."

Bob has some very thoughtful reflections on his maturation as a person and a leader through the years: "The key to one's growth is analogous to metal. You meld it, you forge it, and then you temper it. Let me explain my journey. As a young Ranger coming out of the Army, I don't think I was tempered. It took me three years to decompress. I was just so intense coming out of there that I had a lot of learning to do, or a lot of tempering to get done. I was just too immature still to really figure it out. And it took me a while to figure it out.

"After I started community college, I was getting straight A's. I became president of the student council and president of a business club. I competed in state and national competitions and marketing and won them. I was pretty intense. But as time went on, especially later in my career, I had to learn how to temper all that. I mean, you can take metal, make it hard, but unless you know how to properly apply it, you might do more damage than good.

"And so that tempering of Bob Williams as a leader, as a contributor to my unit and society, took experience. It took mentors who pounded me down when I needed it and built me up when I needed it to give me the right edge to be an effective person or an effective leader. It took a lot of experience to do that and took a lot of maturing also. And much of it was self-reflection and friends providing feedback to me. That goes back to what I found in the 2nd Ranger Battalion: cultivating friendships, cultivating trust, building a team, going through a little bit of fire and ice to get to where you need to go."

Bob's children continued a legacy of service to our nation with both of his sons serving in the military.

His thoughts on military service are shared by so many of his fellow veterans: "I still believe the military is a great place to pursue opportunity, develop oneself, and grow. Sacrifice and even death are always a threat, but it is the same way in everyday life. Living to tomorrow, riches, success, and a happy life are never guaranteed in this life. The values we are taught in the military become our compass in decisions we make tomorrow. The friends we make, the relationships we value, and decisions we make all influence our destiny… whatever that may be. When making the decision to join the military, decide what you want, be the best you can be, and be able to live with your decision."

* * *

Brian Quinlan was raised in Methuen, Massachusetts in a traditional Irish Catholic family. At the age of nine, tragically his father passed away. This crucible forced Brian and his two brothers to work hard at every endeavor and be disciplined in their approach to their responsibilities. In high school, he went to trade school in metal fabrication and welding and truly enjoyed it, but he had an interest in military service that was inspired by his father who had served in the Korean War, and later as a member of the organization of Strategic Headquarters Allied Powers Europe (SHAPE) in France, and a Special Forces uncle that had served three tours in Vietnam.

Upon graduation, he and a friend, Jim German, decided they wanted some adventure, so they went to the local Navy recruiting station to enlist to become Navy SEALS, however, they were told they were too young, and so they then looked at other options. The Army Ranger option appealed to them for a couple reasons: they would be physically and mentally challenged, and one of the original founders of the Rangers was Major Robert Rogers who was born in Methuen, Massachusetts in the 1700s.

The Methuen high school athletic teams proudly retain the name "Rangers." During the French and Indian War, Rogers raised and commanded the famous Rogers' Rangers who were trained for raiding and close combat behind enemy lines. Rogers Ranger Creed contains twenty standing orders/principles for Rangers that still apply today.

Brian and Jim German enlisted for the Ranger option and successfully completed Ranger School as teenagers with no active-duty experience other than basic and advanced individual training and Airborne school. This was quite an achievement at such a young age. Brian's demonstrated drive, attention to detail, physical toughness, and ability to listen and learn would be traits that would serve him well as a young Ranger and later a professional in an extremely technical field.

Brian thrived in the battalion during a very transformative period immediately after its formation. He was a young noncommissioned officer who ensured he and others were trained based on exacting tasks, conditions, and standards. He was blessed to be coached and trained by his squad leader Staff Sergeant Kim Maxin, Sergeant Tom Gould, Platoon Sergeant Roy Smith, and Lieutenant Eldon Bargewell.

He also learned and flourished alongside fellow professionals Fred Kleibacker, John Brasher, Jim Smith, and Gary Longenecker. When he looks back on what he learned in his three years in the 2nd Rangers, he says: "I think we all grew up in an environment where, as young men, we did what were told. And so, I find that in all of the training I received, it did not matter whether it was the challenging and dangerous Ranger School, Jumpmaster training, amphibious reconnaissance, or demolitions training, if you put your trust in your experienced trainers, and you did exactly what they said, you would not only be just safe, but you would succeed.

"I also learned how important mentors and trainers are in your development. I found many parallels between the military and the civilian professions where typically your first priority is safety at whatever you're doing, and then protection of your assets, whether it be the military hardware or your items and equipment that you're using in the civilian job. I also carried forward the importance of detailed planning and documentation in project management and the value of studying lessons learned in your chosen profession."

Brian went on to also talk about the importance of laying the foundation with quality people in your organization: "It was absolutely amazing the high-quality soldiers and leaders that were present during the formation of the 2nd Ranger Battalion. There's no doubt that we were extremely fortunate. But I don't think that it was an accident, and I don't think that it was luck that we found ourselves in those positions.

"When a project or business Team is assembled, whether it is a Squad, Platoon or Company (Civilian or Military), a solid foundation must be laid. Obviously, someone had selected our Ranger unit leaders. The Leaders and mentors that were put in place during the formation of the 2nd Ranger Battalion, became the cornerstones or pillars which resulted in tremendous unit and individual success in the future."

Brian left the military after four years and returned to Massachusetts.

He was enrolled at a local college, but with his previous history of welding, he chose to work as a pipe fitter at a nearby nuclear plant under construction. So, with a high school diploma and four years as an Army Ranger under his belt, Brian went to work helping build and maintain nuclear power plants.

He soon demonstrated for others that he was disciplined, a quick learner, and a hard worker. He quickly advanced in his new profession because he focused on being truly competent in his new profession. He assisted a friend and five other like-minded individuals in the development of a nuclear power plant inspection company and eventually rose to become vice president.

In typical Ranger fashion, Brian stepped up to the plate and went out to various customers and pulled together as much of an understanding as he could. As he states: "When you construct something, when you build something, then you must understand it from the ground up. You must understand everything from the foundation to the top. Nuclear power plants are very complex buildings. I worked hard to learn every technical aspect of the business. It was just like in the battalion. We were serious about what we did. We wanted to do things correctly. We wanted to do things "by the book" of course and follow the chain of command or organizational structure and be true to whatever activity or action was required."

As a civilian leader in a very technical field, he constantly reminded himself of what he learned in the Army: "I always found myself in a position where I needed to do things by the book. We needed to follow instructions and prescribed procedures. Leaders do things correctly while setting the example for others."

This focus on detail and following procedures has resulted in tremendous professional success for Brian. Brian loyally supported his friend for nineteen years, and then moved on to work at various other positions in the commercial nuclear industry. Today, over forty years later, his knowledge and skill in the business is shared as a project manager and consultant.

Brian shares the following behaviors he learned in the military that he applied successfully in his civilian profession:

Know your job and do not cut corners.

Stick to the task at hand and do not quit.

Set the example for others and treat them with dignity and respect.

Be loyal upward and downward.

Embrace mentor advice, and patiently mentor others.

From an expert in small unit tactics to a professional in the commercial nuclear industry, Brian Quinlan demonstrates for all of us that gaining true competence in any field takes intense study, experience, diligence, and self-discipline. Behaviors instilled in him in the service of our nation.

Brain Quinlan's journey from being a young airborne Ranger infantry-man to the builder of nuclear plants is fascinating and speaks to his thirst for knowledge, high standards, attention to detail, and hard work. It is truly a story of living a life of significance.

* * *

Rangers Andy Pancho and CC Lewis decided to stay in the Army after the 2nd Rangers. They each proudly served over 20 years, attaining senior NCO rank. Throughout their military careers, they served in tactical and training units focused on improving the competence and confidence of the soldiers they had the honor to lead. In their own words, here are their stories and the positive impact their time in Bravo company had on them as a person and military leader.

Andy Pancho: "I was born and raised in Hawaii. I come from a family of nine, five sisters and three brothers. I grew up playing sports. Three of my older sisters married service members. Two were in the Army and one was in the Air Force. I always enjoyed hearing of their adventures during their time in the military. I thought it would be something that I wanted to be a part of.

"When I was a senior in high school, I signed up for the delayed entry program. After I graduated from high school, I went to basic training at

Fort Ord, California. My first assignment was as a young infantryman in 2nd Battalion, 2nd Infantry in the 9th Infantry Division at Fort Lewis, Washington. I enjoyed that assignment because we were training all the time.

"One day my squad leader who served with Special Forces in Vietnam and was Ranger qualified, sat myself and another young soldier down and encouraged us to volunteer for the Ranger company that was based at Fort Lewis. We asked questions about what the Ranger company did and how they operated. We both liked what we heard and signed up to try out for positions that were available. We both passed the selection process and accepted the assignment to Bravo Company Rangers.

"Within two months after I arrived in Bravo Company, I went to Airborne School and returned to go through training to become part of the team to which I was assigned. I served there until the 2nd Ranger Battalion was stood up and Bravo Company was deactivated. A few of the Rangers in the platoon I was assigned volunteered for reassignment to the 2nd Ranger Battalion. Tim Grezelka, Al Kovacik, Joe Picanco, and I were assigned to B Company while others were assigned to the other companies. It was a thrilling moment for me when I participated in the activation ceremony of the battalion.

"I went to Ranger School in January 1975 with about 21 other soldiers from throughout the battalion. This was the first group of soldiers from the battalion to be sent to Ranger School. If my memory serves me correctly, we all graduated.

"During the next year as a young sergeant team leader, I was coached and trained by the very best: Staff Sergeant Robert Demoisey, my squad leader, and Lieutenant Bill Leszczynski, my platoon leader. Robert Demoisey patiently gave us all the knowledge he had. He taught us how to be leaders and what he expected of us as team leaders. I had very little practical experience being a team leader or any type of leader, except for what I had learned in Ranger School. I was kind of intimidated because I was a recently promoted sergeant (E5) and the other team leaders had a lot more experience and time in their positions. I knew I had to catch up in gaining experience.

"Sergeant Demoisey acknowledged my inexperience but recognized my ability to listen, accept feedback, coaching, and to quickly learn how to be a leader. He was the most influential NCO role model I had in my career. Lieutenant Leszcynski set clear expectations and then let the team leaders and squad leaders do their jobs. When we did not meet the standard, he would pull us to the side and talk to us in private. I remember I got talked to a few times by him, and it was an ass chewing, without being an ass chewing. And that's what I appreciated about Lieutenant Leszcynski. I learned from him how to truly counsel soldiers without demeaning them.

"I learned the same from Captain Magruder. He was always a calm, cool, collected leader. I never saw him get angry. I never saw him raise his voice. I learned from Bill Leszcynski and Lawson Magruder that by not raising your voice and yelling at people, you could get more things done, and people would listen to you more, and they would tend to absorb what you told them. We trained very hard for the next year until the battalion was certified in December of 1975.

"When I reflect on the values that were instilled in me in Bravo Company, four come to the forefront: Duty, Integrity, Courage, and Loyalty.

"Your duty was to do every task to the best of your ability. It was up to me to learn to do it to the best of my ability so I wouldn't be a liability to the rest of the team.

"And integrity was part of that, because if you didn't know what had to be done or you were not sure of how to accomplish it, I had to either find out by asking somebody else and then getting the training to be able to accomplish the task. That meant I had to have the courage to ask if I didn't know it, and to ask for help when I needed it.

"I also needed the courage to tell somebody if he was not meeting the standard and help him learn and grow.

"I also learned to be loyal to my soldiers. They could confide in me if there was something wrong in their personal or professional lives. I was there not only as the person to lead them through getting the mission or task accomplished, but also as a person that they could talk to. I would stand behind them if they were correct in what they were trying to do. If they were off course, they could count on me to put them on the right track. They needed to know I was competent and knew what I was doing, and when I interacted with them, they knew that they could trust that I wasn't going to lead them improperly. They could trust me."

Andy absorbed those leadership lessons from Bravo Company and paid them forward throughout the rest of his career. He became the consummate "Be, Know, Do" NCO. His positive leadership skills set the example for thousands of soldiers and fellow leaders over the next 20 years. He was a trainer focused on developing competent soldiers and junior leaders in a variety of tactical and training units. He was an infantry drill sergeant, a Darby phase and Desert phase Ranger instructor, and a Long-Range Surveillance Course instructor.

He had a second tour in 2nd Ranger Battalion, this time in Charlie Company as a Squad Leader, then platoon sergeant. He then moved across the country to Fort Drum, New York to serve as a first sergeant and operations sergeant in the 3d Battalion 14th Infantry (Golden Dragons) in the 10th Mountain Division where he was reunited with Brigadier General Lawson Magruder who was his commanding general on Operation Restore Hope in Somalia.

After his return from Somalia, Andy was assigned to the 6th Infantry Division in Alaska. He again asked to be with soldiers and found himself as a first sergeant in 1st Battalion, 17th Infantry in the 6th Infantry Division at Fort Wainwright outside of Fairbanks. He was in his comfort zone training arctic light infantrymen when shortly after assignment, he was faced with a career decision. He was selected to attend the Sergeant Majors Academy at Fort Bliss, Texas, which would ensure his promotion to sergeant major. However, Andy made a family decision to decline attendance because it would have meant multiple high schools for his daughters.

The family also loved Alaska, "the Last Frontier." He retired in 1995 after 21 years of very honorable service to the nation.

Andy and his wife, Jerenda, have lived in Alaska for over 30 years. He worked as a maintenance mechanic for the local post office for thirteen years and then Plant Operator for a local medical clinic, also for thirteen years. In both positions, Andy was recognized by his supervisors and fellow employees for his intense loyalty to the organization, his focus on fixing a maintenance problem with a lasting solution, his thirst for knowledge, his can do, positive attitude, and his self-discipline and work ethic. His focus on competence was a shining example for others, to include his family. His daughter and son followed in Andy's boots and proudly served in the Army's Military Police Corps and his grandson served in the Army National Guard.

* * *

CC "Cliff" Lewis was one of a handful of original Bravo Company plank holders who retired as a sergeant major (E9), the highest enlisted rank in the Army. Only 1% of all soldiers attain the rank of sergeant major. His career spanned from 1974 to 2000. Over those three decades, he helped his beloved Army transform into the highly trained and ready force that won the Cold War and soundly defeated the Iraqi Army in Desert Storm. When he could have chosen most any other profession, he elected to stay in and help train and lead soldiers in a variety of units. Rather than seek monetary fortune, he decided to fuel his purpose to serve others as an extraordinary military leader.

From CC Lewis: "I was raised in St. Charles, Missouri. I joined right out of high school because of my father, although he said time again and again not to join because of him. And I said again and again, "Oh, I'm not doing it because you did." He enlisted in 1949 and was on Okinawa when Korea started. His 76th Engineer Construction Battalion was one of the first intact units to land in Pusan. They "parked their equipment and went into the line as Infantry." They faced mountains, a freezing winter, and chaos, but ultimately like all soldiers they got what they had to get done.

"Hearing his stories about soldiers and soldiering and watching the movie about the Green Berets with John Wayne, it became obvious to me the Army sounded more and more like I should participate, so I enlisted. It was funny that I had originally enlisted to be an engineer in the 82nd. But when it came time to sign my enlistment paper, there was a glitch, and I couldn't be an engineer. So, I told the recruiter I want to be a Green Beret. He said we can't do that right now, but we could put you in Ranger School and you know most Green Berets are Ranger qualified. So, I took the Ranger option.

"After graduating from Ranger School, I traveled to my first duty station at Fort Lewis and was assigned to the 3d platoon of Bravo Company. I spent the next five years growing up in Bravo Company. I went from being an E5 team leader to being an E6 platoon sergeant. My years in Bravo Company were extremely impactful and laid the foundation for my success as a leader."

"**Here were my big takeaways from the experience:** The first had to do with gaining competence and confidence through training. I went through a phase early on in my first couple of months of doubting myself. Many of our leaders were pretty intimidating when it comes to combat experience. So many had served in Vietnam: Captain Magruder, First Sergeant Block, Lieutenant Bargewell, and all the platoon sergeants. For a young leader like me, that was intimidating and worrisome. I mean, I knew enough from my dad that combat, if we made the trip, was going to be a big deal.

"At some point I ended up checking in with Platoon Sergeant Smith and saying to him, I was worried, wondering about combat and how I would do. His answer was succinct: He said, "Look, you do what we tell you. You train hard, you'll do okay." His guidance probably lasted longer than that, but that's how it's been summarized in my head for over 20 years. It became kind of my mantra to younger folks as I went on: "Don't worry too much about combat. Instead, focus on training. Focus on meeting the standard but do what we say. Do it the way we teach it, do what you need to do, and then when you get off to combat, it'll work out."

"We learned daily the true definition of Duty. We had a job to do, and we were going to do it. And nothing was going to get in the way. We would joke about "the jump in" and how it "would be a one-way trip" but "that would be ok." Gallows humor, maybe, but I don't doubt any of us wouldn't have gone when the green light came on.

"A quick story about senior leader presence and sharing hardships: During the battalion's first jungle training rotation to Panama, our squad was the rearmost unit in the company on a night march through the jungle. Immediately behind us was the battalion tactical command post. Somewhere around 1:00AM, I looked back and there was a Ranger as beat down as any of us but clearly humping a second rucksack on top of his own rucksack. I did a double take and kept moving. I looked back later and realized it was our battalion commander, LTC Bo Baker. The battalion headquarters had a Ranger who was ailing, and they had split up his load and were passing the ruck around. I don't remember my exact thoughts at the moment but later they coalesced into "what a great example."

"I saw senior Leaders lead the way out of the aircraft during airborne operations. No matter how bad it got, a senior leader was always there leading the way."

"There was a calm, cool, and collected manner in the leadership environment in Bravo Company. I don't remember a lot of yelling. I'm not sure we achieved what was later described as "Quiet Professional", but I don't remember us running around with our heads cut off. Instead, whatever the task, let's get it done even under the most miserable conditions."

"Training management was emphasized at every turn, and we truly understood it. We had tasks, conditions, and standards and understood the difference between performance-oriented training and lectures. At the most basic of levels, we used the word "Standard," and meant it and lived by it. That was extraordinary and clearly valuable.

"The six months of intense training leading up to our successful battalion certification exercise in December 1975, would be the model I recommended to my senior leaders in the future. We went from Individual, then Team, then Squad, then Platoon, then Company, and finally a couple of weeks of Battalion training. That emphasis on small units paid off in the long run, because of the firm foundation it laid for the battalion. When I would describe that sequence and emphasize that there was only one battalion field training exercise over just a couple of days, NCOs and Officers marveled because it wasn't what they'd seen in the rest of the Army. What they'd seen was battalion level operations ad nauseum.

"I watched and learned from exceptional leaders who were great trainers and men of high character. My first squad leaders were Sergeant Al Kovacik and Sergeant Jimmy Bynum. Both were smart, fit, and good trainers. My platoon leader was Lieutenant Bargewell. He was an excellent trainer and by that, I mean he did a great job of walking the fine line between individual training and team, squad, and platoon training while keeping an eye on the eventual requirement for company and higher. There was a very enviable and note-worthy air about "the Barge" which I know I never achieved, but it was my goal.

"I learned the meaning of competence. I remember when I started getting ready for the E5 promotion board, one of the answers I had to memorize were the four indicators of leadership. They were morale, esprit de corps, discipline, and proficiency. Over the years, I figured out that morale and esprit were very hard to quantify. I mean, you knew it if you saw it, but it was hard to put a number to them. On the other hand, discipline and proficiency were easily quantifiable. They matched up, as I discovered through my Ranger training with standards.

"And then a revelation hit me because the original question was what are the four indicators of leadership? Where's good leadership? I realized that where there were units of high discipline and high proficiency, morale was by definition, high. Always. On the other hand, where you would find low morale and esprit de corps, you would also find low discipline and low proficiency. And that kind of became a tool I used for the rest of my career.

If I would emphasize what I could control, discipline and proficiency, morale and esprit de corps would fall in line and come along."

Cliff decided to remain in the Army and served for another 19 years. He was a shining example of General Abrams original charter for the Ranger Battalions where he wanted the officers and NCOs to go from the battalions to other units to help transform and make the Army better wherever they went. He opted to be in a variety of tactical and training units, seeking assignments where he could positively influence the next generation of soldiers. He was a drill sergeant, a scout platoon sergeant, a Ranger instructor, a first sergeant in the Ranger Training Brigade, a staff sergeant major, a Drill Sergeant School Commandant, Command Sergeant Major for 2nd Battalion 19th Infantry, a basic training battalion, and Command Sergeant Major for 1st Battalion 503d Infantry in Korea.

In each of these postings, Cliff had a positive influence on others. With his expertise and experience, and an enthusiastic booming voice, he inspired young soldiers and leaders to seek proficiency with a positive, can-do Ranger spirit.

Cliff and his wife Sherry Ann decided it was time for Cliff to retire in 2000. They settled in Cuba, Missouri. Cliff says, "We live out in the country and heat with wood. So, I spend a lot of time cutting down trees and splitting them! We have made trips out to my son's in Arizona and my daughter, who has lived in couple of places in Europe and Kansas. We spend a lot of time with our grandchildren."

For the past 25 years, Cliff and Sherry have been members of a small Baptist faith community in their hometown of Cuba. Cliff is a deacon in the church. It is a church that emphasizes the preaching of the Word. Cliff says when he preaches, it is not adding to or taking away from what God said in the Bible. He just tries to preach what God said without inserting his personal opinion. The small congregation of 50 people are worried about folks' souls and getting people saved. They are a very loyal flock.

Cliff said that about ten years ago his church family was rocked when it's pastor of many years suddenly departed under questionable circumstances. Just when the congregation could have collapsed, Cliff stepped forward in typical Ranger fashion to steady the family. "I emphasized to the folks right away the need to keep everybody informed so that rumors don't get passed around and we need to trust God. Or to compare that to my Army time, you need to just trust the chain of command.

"In this example, with the pastor moving on, we need to trust God. He's going to provide. We must be able to actively and visibly trust God. You simply trust him and get on with your day. We're going to trust him to provide the new pastor, and it'll be okay. As it turned out, God provided us with an excellent man who then served for ten years. It was a real blessing. But a church our size could have easily fallen apart in that kind of situation but didn't."

This speaks to the determination Cliff instilled in them and his positive leadership style.

We will close this chapter with inspirational words from Cliff Lewis about why young people should consider serving in our military: "I do get the chance to talk specifically about military service with teenage members of my church. You can almost see the dread in their eye when they get to church and I come around the corner, or they come around the corner and I'm standing there, because they know I'm going to brace them up against the wall, figuratively speaking, and quiz them on how church went, how school went, what they learned. And then from there, at some point in their late sophomore or junior year, I start hinting at and tracking down the parents and saying to them that they need to get their son or daughter down to talk to a recruiter. I phrase it in the context that it's obvious that they should at least see a recruiter and talk to them, because I think it is obvious Americans should want to serve.

"I don't push that we should return to a draft. But the idea of service is something I will passionately share with young people. I do briefly mention the benefits like pay, medical, and education, but most of the conversation is about service to something greater than self. "Hopefully, they'll walk away from me convinced that I think that service to the nation and defense of the nation is an important thing, and by participating in the United States military, they can have a piece of that. They will also get to serve in some wonderful places and gain a skill that is marketable on the outside. I'll use those side issues as a tool, but I won't let it distract me from the overall point, which is, it's an honor to serve. It was an honor for me to serve. It was an honor for me to be a soldier."

"Individual competence will lead to individual confidence, unit competence, and ultimately mission success."

— CPT Lawson Magruder

9. EXPANDED WORLD VIEW: RESPECT FOR OTHERS

Training and real-world deployments will remove your personal blinders
and help you gain respect for others and understand the immense role
our military has in demonstrating the principles of democracy for the rest
of the world.

—CPT Lawson Magruder

From left: Jim Dubik, Bill Leszczynski

JUNE 13, 2007, SAMARRA, IRAQ: THE SECOND BOMBING BY AL QAEDA SUN-
ni insurgents had again severely damaged the Al-Askari Mosque. The
purpose was to reinvigorate the Sunni/Shia sectarian divide that fueled
the insurgency. Al Qaeda was afraid that with the US surge, they may
not get another chance. It didn't work, for a variety of reasons, but in the
flurry of activity, the Iraq Minister of Defense called LTG Jim Dubik, the
Commanding General of the Multinational Security Transition Command-Iraq (MNSTC-I) and the NATO Training Mission-Iraq (NTM-I).
He asked Jim if he could borrow a helicopter to fly to Samarra from Bagh-
dad. He needed to be there immediately.

Jim recounted the event: "I said, no, I won't lend you a helicopter. I'll
take you to Samarra; we'll go there together. My thinking was that if I

went with the Minister and we got in trouble, a rescue team would come and get us both. Sharing the danger with the Minister allowed us to bond instantly. The conditions in Samarra at the time were extremely dangerous. When we landed, an 82nd Airborne Division unit provided security. As we were driving down the street toward the mosque, everyone was shuttering their shops and closing their windows. The streets were vacant. It was an eerie feeling.

"While we were at the mosque, we visited several Iraqi military units responsible for security of the site. When we were getting ready to come back, the Iraqis provided the security to our helicopter. The minister told one unit to go back the same way we came. And then he directed another convoy to go back a different way with the two of us. The first convoy was attacked, and we were not. From that day forward for the next 14 months, the Minister and I had relationship built on trust and shared risk."

Jin Dubik's quick thinking that day was a result of four decades studying philosophy, war and peace, and human behavior, and leading in a variety of worldwide operations prior to Iraq. His expanded worldview and respect for others started at an incredibly young age. He nurtured both through hard work and constant study and learning.

Jim was inspired to join the military by his uncles and father who were World War II veterans. As Jim related: "They all talked about their service, and that was a source of inspiration. I went into seminary in my freshman year of college with the intent to become a Catholic priest. But I left the seminary in 1968 when the USS Pueblo was captured. I figured, well, we're going to war, and I should join up. (One indicator that I may have been called but was not chosen.) So, I left the seminary and went into Army ROTC at Gannon University in my hometown of Erie, Pennsylvania, . It was mandatory at the time. I had really no intention of making it a career; I had intended to enlist in 1968. My dad said, 'Son, don't do that now. Go finish your degree. Go in as an officer and then serve.' So, that's what I did.

"After I graduated, I had an opportunity for a regular army or reserve commission. And I was leaning toward the reserve commission because it was only a two-year commitment. I asked my dad about it, and he said, 'So what's your degree in?' And I said, 'Philosophy.' And he said, 'And what job do you have?' And I said, 'None.' He said, 'Three years sounds like a good plan for you.' I accepted the regular Army commission.

"His rationale was that with a reserve commission, I would have to ask to extend, and depending on the size of the Army and what would happen when the Vietnam War ended, I may or may not be able to stay in. He was in World War II and Korea, and he saw the shrinking of Armies afterwards." Following his Dad's advice began a journey in the Army for Jim that would last over 37 years.

Jim was commissioned a regular army officer in the Infantry in 1971 and was immediately assigned to the 82nd Airborne Division, ending up in B Company 3/325 Infantry. After serving there for several months, he attended the Infantry Officer Basic Course, Airborne School, and Ranger School. Upon his return to the 82nd, he ended up first as the Scout Platoon Leader in HHC, then to Company C as the Executive officer. During that tour, he would start a friendship with then Captain Lawson Magruder, Commander of Charlie Company. They would both join the 2nd Ranger Battalion two years later as plank holders.

Jim spent four years in the 2nd Rangers. He had various staff positions in the battalion and was honored to be the first non-former company commander to be selected as Lawson's successor in Bravo Company. This was quite an achievement considering the vast combat and tactical experience the senior captains in the battalion had. But Colonel Bo Baker, the battalion commander, and his executive officer Major Bill Powell recognized Jim had the tactical and training moxie, along with the mature, competent leadership abilities that were necessary to command a company of outstanding professionals.

Jim flourished and grew during his time in command and staff positions in the battalion. In his own words, "What I learned in the battalion carried with me throughout my military career. There were multiple role models for me in the battalion," he added, "some of the role models were people, others were groups. And in both cases, they were people that I wanted to be like. They were very stimulating, interesting, and fun to be around. I'll start with the groups first.

"The sergeants were amazing, absolutely amazing, and inspirational in their capacity, in their desire to do well. This is 1975. I mean, you know what kind of shape the Army was in at the time.

"And here were a group of mostly Vietnam veterans or recently promoted sergeants who were wanting to create something for the Army that they knew would be better. It was just inspirational to see what they could do and how they could do it. I carried that with me the rest of my career, making sure that I always gave NCOs the space to do what I knew they could if they were empowered to do so.

"First Sergeant Bruce Pross and Roy D. Smith were senior NCOs who were very formational in how I approached my job as an officer, not just as a company commander. They helped me succeed as a company commander, but they also taught me so much about the officer and NCO relationship and about what to expect from sergeants and soldiers. When I made brigadier general, I sewed sergeant stripes behind the Velcro that held my rank, just to remember how important they were in my own development.

"The other group, the officers. Generally, though not universally, because we know we had a couple of characters," Jim continued, "the officers were all examples of people who took their profession seriously. All of us had the necessary physical capacities, which is no small thing, and I don't mean to diminish that. But beyond that, the officers were, for me, inspirational in how seriously they studied, and they took the profession. Leaders like Dave McMillan… he just knew and read so much and knew how the battalion fit into the profession and how the profession fit into the nation, and he was just one example.

"Bill Powell, Buck Kernan, John Abizaid, Terry Fullerton, Len Fullenkamp, and Lawson Magruder were other examples — to name just a few."

Bill Powell taught everyone skills of how to run a staff that Dubik kept with him forever. Bo Baker and Wayne Downing taught him things about leadership and aligning your subordinates with the main mission and decentralized command. During this early part of his career, Jim found it awe inspiring to be around a group like that — each person competing to be his best and contributing to the rest. What Jim learned, however, was well beyond the technical professional aspects of the profession.

He explained, "Before I assumed company command, I did not have the benefit of going to the Infantry Officer Advanced Course, which is focused primarily on company command." He then related this story which speaks to the importance of understanding all aspects of your profession, trust in your teammates, and how to provide professional feedback.

"The special trust formed within the battalion started with Colonel Bo Baker. He set the tone in the way he treated everyone with equal respect, regardless of who you were or what you were doing. That is how he led and taught. For example, I wrongly gave an Article 15 (nonjudicial punishment) once, wrongly. I can't remember many of the details, except that whoever came to see me convinced me that this young Ranger needed to be punished for his misconduct and deserved an Article 15. After I did a cursory investigation, I found the Ranger guilty and meted out punishment.

"Within his rights, the Ranger appealed to the battalion level. After his review, Colonel Baker overturned the Article 15 punishment and called me to his office. He sat me down and said, in a fatherly way, in a completely developmental way: 'Look we run a justice system for the Army. We don't run a kangaroo court. And you're responsible as an officer with the authority, you're responsible for using that authority fairly, and you did not in this case. I don't want to see this again.' But he didn't chew me out. He didn't make me feel belittled in any way, and that builds a lot of trust. I didn't walk away afraid of making a mistake. I walked away saying, okay, well, I have something to learn here, which I certainly did. And that kind of example was top down, and lateral as well."

When discussing the positive learning environment and values found in the battalion, Jim had this to offer: "Everyone — from the Ranger private to the battalion commander—co-created the environment,. It wasn't that one or more of the values came to the forefront. They all came to the fore at one point or another in any given week. And the same thing with leadership attributes. Every week, there would be some examples in which either a Ranger, a sergeant, a peer, or a senior would provide an example of deep character and the requirement to be present. Leadership is an activity of presence, and every day that was reinforced by every leader.

"For me, again, it wasn't any single value. It was the fact that before the Army formally adopted the seven Army values, they were present in 2/75. And before I learned any doctrine about leadership attributes or competencies, I saw them in action in 2/75. Values and leadership were a given. And when you're in that kind of environment, it draws out the best from you because you want to present your best to those people who are presenting their best to you. It's the mutuality of the whole kind of experience that for me is much more important than any single one. "

On relationships he stated: "There was never a time where one of my peers or a senior or one of my subordinates would not have the time for an interaction. If I went to a person and had a conversation, whatever the conversation was about, there were never anyone who turned away and said, 'I don't have time for this.' It was everybody contributing to everybody else."

As Jim's tour in the battalion wound down, he made a key career decision to stay in the Army. "I realized I wanted to stay in the Army. Up until this decision, I was motivated externally. And by that, I mean I loved jumping out of airplanes. So, every time I jumped out of an airplane, I was motivated to jump out of another one. I loved doing challenging physical training. I liked all that physical stuff. I liked going to the field. I was motivated by external things.

"But after company command and reflecting on it, I became internally motivated that I didn't need something or somebody to motivate me to do well. Something happened in company command that said to me, this is a profession worth staying in for. There is a contribution you can make. You can maximize your potential. You can maximize your talents in the service of a noble cause. You can do things that are beyond yourself in this profession. This profession has longevity."

And Jim's profession clearly had longevity! His next two decades is a story of a search for knowledge, demonstrating humility, and facing daunting personal and professional challenges with faith, courage, and conviction. It is a story worthy of a separate memoir that we hope he writes someday. But for this chapter, we will highlight just a few of the turning points in his life story worthy of sharing because the lessons are so profound and speak to the shaping of his expanded world view and his deep respect for others.

As Lawson wrote in his book, *A Soldier's Journey Living His Why*, Jim was a lifetime learner. Throughout his adult life, Jim has had a thirst for knowledge. He actually brought his professional books into his office and ran a lending library to anyone interested. When he wasn't studying his military profession, in his spare time, he continued his study of human nature, the meaning of mankind, and philosophy.

As a young major, he attended the amphibious warfare school in Quantico, Virginia, which was followed by a master's program in Philosophy at Johns Hopkins University, Command and General Staff College at Fort Leavenworth, and teaching for three years at the United States Military Academy. (Of note, he had all but his dissertation completed for a PhD after only two years. But a professor at the University blocked his progress because he thought Jim had completed the requirements too quickly. A couple of years after Jim retired from the Army, he went back to Johns Hopkins and completed the requirements for his PhD!)

After battalion command, rather than attending one of the senior service colleges, Jim was handpicked among select candidates to attend the U.S. Army's School for Advanced Military Studies (SAMS) Fellowship program — a year of very concentrated study of the operational level of warfare, followed by a second year teaching majors.

That was the last formal military school that Jim attended, but his thirst for knowledge continued after the Army. His doctoral thesis resulted in him authoring *Just War Reconsidered: Strategy, Ethics, and Theory*.

He is currently a senior fellow at the Institute for the Study of War, teaching in the Hertog War Studies Program and had been a professor at Georgetown University's Security Studies Program for several years. He also has published over 200 essays, monographs, and opeds. Additionally, he's written a number of chapters and introductions to books and coauthored *Envisioning Future Warfare*, with the late General Gordon S. Sullivan. His teaching and writing resulted in him being recognized as a thought leader within the Army, for the armed forces, and in the wider U.S. national security arena. In 2024, the Army established the LTG James M. Dubik Writing Fellowship for the Study of the Military Profession.

Jim was a student of humble leadership as a young officer. Later in his career he modeled it for so many others. Here are some of his insights on the topic: "Humility is an especially important leadership attribute. It becomes more important the more senior you get. As a junior officer, whether as a platoon leader or company commander, or even as a battalion commander, you can become so knowledgeable of that organization that you may think you don't have to be very humble. Because your experience so outbalances that of your subordinates, you may think you know the answer to most questions.

"But once you get past battalion command, you start being in a situation where not only do you not have the answers, but you also don't even understand the problem. That's when you must be humble to listen to other people who understand things you don't. And that's when a person without adequate humility starts to derail."

Jim shared the story of his first assignment as a new brigadier general. He was assigned to the Army Staff as the Director of Training. He was totally out of his comfort zone: "I oversaw $10 billion of the Army's budget. With my philosophy degree in hand, I told the staff when I took over, I can tell you what a number is. I just don't know what to do with it. And so, I asked that they educate me for a year. It was a powerful education that I received at the hand of one GS 15 DA civilian, a colonel, and a couple of other civilian and military staff, on how the budget runs, how it works, how the training budget fits into the operations budget, which fits into the Army budget.

"Because I admitted I was not knowledgeable about the budget process, my team helped me, and thereby the whole team, to succeed during that year. My willingness to learn goes back to my early days. Humble leaders know what they don't know."

Jim, like the rest of us, faced many daunting challenges and crucibles in his life. He handled each of them in a resilient, adaptable, and courageous manner, supported by his deep Catholic faith and strong relationships. Here are just a few of them and how he either handled the challenge or the crucible:

He overcame two mediocre officer efficiency reports as a major by excelling in subsequent assignments.

He went through a difficult and troubling divorce after 20 years of marriage, just as he went into brigade command. His strong faith, his daughters and family, and his friends helped him through this period.

As a brigade commander in the 10th Mountain Division on a humanitarian mission to Haiti, he and his Brigade Combat Team (BCT) were the supporting effort, responsible for establishing a safe environment for the return of President Aristide in the Northern portions of Haiti. The BCT was able to accomplish some incredibly challenging tasks together. (And while in Haiti, he started corresponding with a woman he met prior to deployment. The letters led to a now nearly 30-year marriage.)

As the Chief of Staff of the Army's point man for what ultimately became Stryker Brigade Combat Teams, he and his Brigade Coordination Cell were able to develop an innovative approach to training, equipping, and leader development for the brigade. He also helped convince Secretary of Defense Donald Rumsfeld of the requirement for the Stryker vehicle, rather than using an old, mechanized vehicle (M113). The Stryker vehicle and concept proved itself in both Iraq and Afghanistan. His demonstration of moral courage against all odds with a quite resolute, or many would say hardheaded Secretary of Defense, was a model for other senior leaders.

In Jim's culminating career assignment, Commanding General of MNSTC-I and NTM-I, he leaned on his mature world view and training experience that started in the 2nd Rangers, his previous experiences in the Pacific, Haiti, and Bosnia, and strong personal relationships with Generals Ray Odierno and Dave Petraeus to succeed in accelerating growth of the Iraqi Security Forces in size, capability, and confidence. His focus on training lethal squads and platoons led by competent leaders who then partnered with American units contributed to the positive results of the surge in 2007-2008.

Jim has continued to lead the way in retirement. Along with teaching and writing, he has dedicated many hours to a Catholic non-profit, Leadership Roundtable, that seeks to identify, develop, and implement best practices in management and leadership in the Catholic Church in the United States. In true Ranger tradition, Jim continues to Lead the Way!

* * *

Bill Leszczynski was destined to join the military. "My father was a veteran. He was a B-17 waist gunner and was shot down over Czechoslovakia on a bombing mission into Poland. He was a prisoner of war for over nine months. After World War II, he served in the Massachusetts National Guard. Everybody in my family who was not classified "4F" was in the military. I had two uncles who were on the aircraft carrier USS Wasp when it was sunk in the Pacific. Fortunately, they were pulled out of the water and served in the fleet until the end of the war.

"My father never pushed me, or even encouraged me, to go into the military. I was a senior in high school ready to go to the University of Massachusetts. My father came home one day from work and related that a fellow worker suggested I apply to West Point. My father and I discussed it and I applied to the three service academies.

"I had recently seen the movie, *The Longest Day*. I was really impressed with the Airborne and the Ranger scenes in the film. I thought I'd really like to do that someday. West Point was my number one choice. When you fill out the packet, they ask for an alternate. I put down the Air Force Academy as my first alternate. A few months later, I'm sitting around watching the evening news with my father, and the phone rang. It was a captain calling from the Air Force Academy. He introduced himself. He said, "I'm captain so and so. You have been accepted to the United States Air Force Academy. Will you accept the appointment?"

And I said, "Thank you very much, sir, but I'm not interested."

He paused for a few seconds and said, "But you put us down as your second choice."

I said, "Well sir, I really don't want to attend the Air Force Academy. I don't want to go to Annapolis either. But they told me I had to put down a number two choice, so Air Force was my number two choice."

He then asked, "Are you sure you don't want to accept the appointment?"

I responded, "Yes sir, I don't." He thanked me and hung up.

My father was half listening to the conversation, and he asked, "Who was that?" I responded, "Some captain from the Air Force Academy."

He asked, "Well, what did he want?"

I said, "He wanted to know if I wanted to go to the Air Force Academy."

My father yelled, "What did you tell him?"

I said, "I told him no." Of course, my father was absolutely appalled because I didn't have the appointment yet to West Point. Fortunately, a short time later, I got the appointment to West Point. You know, if I didn't get the appointment, I could be painting bridges someplace. Who knows!"

During Bill's four years at West Point, he was positively impacted by many of his Tactical Officers and instructors. But one, Colonel Ralph Puckett, Jr., changed his life forever. Ralph was a Regimental Tactical officer for two years. As an Infantry combat leader veteran from the Korean and Vietnam Wars, he was an incredible role model for Bill and many others. He was the consummate professional who lived West Point's motto: Duty, Honor, Country.

He was also a proud Airborne Ranger who heroically led a Ranger Company in Korea. One of his most important messages to Bill, and the other cadets he coached and mentored, was, "I've never accomplished anything in my life of any significance by myself." Of note, Ralph was belatedly awarded the Medal of Honor in 2021. He passed away in April 2024 during the writing of this book.

Ralph Puckett's example convinced Bill to be commissioned in the Infantry and become an Airborne Ranger. Little did Bill know that the next 28+ year journey in uniform would shape his world view and result in him representing our nation as the Executive Director and Chief Operating Officer for the American Battle Monuments Commission.

Bill's first assignment was as a platoon leader in C Company, 1st Battalion, 504th Parachute Infantry Regiment, 82nd Airborne Division. He was blessed to have Captain Dick Malvesti, another proud leader from Massachusetts, as his company commander. Dick was another Ralph Puckett style leader: competent, caring, and humble.

That personal relationship would continue until Dick's sad death in 1990 on a parachute jump. Footnote: Dick Malvesti would command a company in the 2nd Rangers a year after Bill arrived in the unit.

Bill and his wife Jennifer arrived to the 2nd Rangers in February 1975. After a brief time as a platoon leader in C Company, he was reassigned to Bravo Company where he was a platoon leader, then company executive officer for Lawson Magruder and Jim Dubik. When asked about his big "takeaways" from his time in Bravo Company, Bill highlighted the influence of positive role models who always demonstrated respect in how they worked with others: "We had whole battalion worth of what I considered role models. It was a phenomenal group of young, enlisted Soldiers, NCOs, and officers. But when I think of role models, I immediately think of three. Of course, the first one was Colonel Baker. I don't think we could have ever had a leader who was better for us. He was a perfect leader for the people in the battalion. He set the example for me. I always thought that he was like a grandfather. We always knew what he wanted. You knew when you didn't meet his standard because he would tell you, not in a scolding, but respectful way. He never made you feel bad. He motivated you to get better.

"My second role model was Eldon Bargewell, a fellow platoon leader. 'When you come right down to it, those of us with no combat experience were often lost like a ball in tall weeds. Eldon Bargewell always knew what he was doing. But he was so nice about how he helped you. You could ask him a question and he would never make you feel like you were stupid. He would never, ever roll his eyes. He would respectfully and patiently help you.

"My third role model was Captain Magruder, our company commander. He always seemed to know what you were doing. Whether he did or didn't, he always came across like he really knew what you were doing. I think that inspired us and made us feel confident in what we were doing. I never saw him lose his temper, although I can think of many instances where he certainly could have, and it would have been justified.

"He always treated people with dignity and respect, and he knew how to get the most out of people. He talked to us and didn't lecture us. If he found a problem, he always attacked the problem immediately and fixed it. He never attacked the individual. I think that is very important when you are dealing with young people. He also had quite a sense of humor."

Here's a short story about Lawson's sense of humor: "One of my machine gunners at the time was Specialist Tim Martin, who was part of a three-man gun team. The entire team went out the night before a payday in-ranks inspection and got drunk. The next morning, they are standing in ranks in their Class A uniforms waiting for Captain Magruder to inspect them. All three of them were wobbling back and forth. One of them looked like he was going to pass out. Their ties were askew, and their uniforms were disheveled. They looked like they had just low crawled across the battalion area. They reeked of alcohol and had not shaved.

"We were waiting for Captain Magruder or 1SG Bruce Pross to explode. Captain Magruder very calmly let all three of them know that they hadn't met the standards expected of Bravo Company Rangers. He didn't insult or humiliate them. He never raised his voice. No profanity, but he forcefully said he didn't want to see a repeat performance because he knew that they were better than what they had shown in the inspection. He was right. They were all good Rangers, but they were kids.

"I thought Tim Martin was going to cry. We all walked to the rear of the platoon and Lawson started laughing. I remember Lawson, Bruce Pross, and Platoon Sergeant Taylor all doubled over laughing. Lawson accepted the fact that these guys were young Rangers. In fact, they were young kids and young kids occasionally go out and get into a little bit of trouble. "People think of Tim "Griz" Martin as being this mature Delta Force operator who did such heroic things and made the ultimate sacrifice in Somalia. But once upon a time, he was like the rest of us. He was just a young kid. We never had any problem with those three again. They towed the line and never got into any trouble again. They became the model Rangers for the platoon. The thing that I remember is that Captain Magruder had the opportunity to crush them. He could have insulted and humiliated them to make a point, but he didn't.

"As a result, that life lesson in accountability changed their lives forever. In 1993, my wife and I went to Scott and Mimi Miller's wedding. Tim Martin and his wife were also there. He brought his old *Sua Sponte* Yearbook. One of the stories he proudly recounted to everyone was about the infamous inspection in ranks."

When Bill discussed the values that were imbued in him during his time in the battalion, the one that stood out was courage, in particular moral courage. "I learned the true meaning of moral courage in the battalion, specifically telling your boss what he didn't want to hear, even when he didn't ask. I saw so many great examples where officers and NCOs would stand up and candidly state their opinions. They would stand their ground. They would battle right up until the last minute. "Then when the decision was made, they would salute and they would execute, just like it was their own personal decision.

"But the thing that is burned in my mind is that you've got an obligation to do that, and you've got an obligation that when you're asked a question, to respectfully tell people what you candidly think. And even if you're not asked to provide feedback, you have the responsibility to tell people what's on your mind, especially when it affects your Soldiers."

When asked about how to build trust and a winning spirit, Bill had this to say: "The entire leadership team was with you all the time, and when things were bad, everybody was out there. The company commander and first sergeant were always there experiencing the same hardships. When we were miserable training down in South Rainier, the toughest training area on Fort Lewis, and we thought we were going to die, you looked around and there was the company commander and the first sergeant. Nobody was sitting back in a barracks.

" I think trust comes from doing difficult things. If you're doing airborne operations or live fire exercises and want to do them well, you need people around you to help and support. I always knew I could count on every officer and NCO in the company. Everybody was in the boat together. We trusted each other because of shared hardships."

Bill was one of eight 2nd Ranger Battalion plank holder officers who went on to become general officers. He served with some in other units. Here's what he said about their shared attributes that were forged in the battalion: "The first thing that stands out is they were totally dedicated to the mission. Whatever it took, they were going to get the job done. Second, they were all selfless. Third, they took care of their people. And finally, they were all stand-up leaders. They were going to let you know what they thought. They weren't going to back down and they would not shirk away from making tough decisions."

Bill took his personal and professional leadership lessons from Bravo Company with him on his incredible journey in the Army. He served in Italy as the XO and Deputy Commander of an airborne battalion combat team, was the XO for 2nd Ranger Battalion and commanded a light Infantry battalion during Operation Just Cause in Panama. As a colonel he commanded JTF-Bravo in Honduras and was privileged to command the 75th Ranger Regiment.

As a brigadier general, he served on the Joint Staff in the Pentagon and was the Assistant Division Commander for Support (ADC-S) for the 82nd Airborne Division. At every stop, Bill was known for his competence and his caring, humble leadership style, and that he was supported by an incredible Army wife, Jennifer.

Bill was recognized for his tremendous senior leadership talent and potential for command with his selection for promotion to major general. However, a higher "calling," a senior position at the American Battle Monuments Commission came his way that he could not turn down. It was a difficult decision, but one that fueled his purpose in life to serve our nation and our fallen Soldiers in a special way.

This opportunity evolved from two trips Bill made to Normandy during his career. "I had been to Normandy on two occasions. The first time I was the XO of 2nd Ranger Battalion. We were training in Europe in 1987. One weekend before we left to return to Fort Lewis, a bus load of the officer and NCO leaders in the battalion took a trip from Bad Tolz, Germany, to Normandy. The trip included stops at Pointe du Hoc, Sainte-Mere-Eglise, both Omaha and Utah Beaches, and the Normandy American Cemetery.

"Clearly the absolute highlight for all of us was Normandy American Cemetery, the final resting place for more than 9,300 Americans, including 46 Rangers. Although the stop there was very brief, I vividly remember the cemetery was in perfect condition. Not a blade of grass was out of place. It reminded me of Augusta National during the week of the Masters Tournament. The second time I went to Normandy, I was the ADC(S) in the 82nd Airborne Division. I took a contingent from Fort Bragg for the D-Day Anniversary Commemorative Celebrations. It was a spectacular trip. This time I got to spend a few hours walking through the cemetery visiting the graves of the Rangers and paratroopers interred there. I also met the Superintendent, an Air Force Veteran, the Assistant Superintendent, also a Veteran, and some of the French staff at the cemetery. They talked about their sacred mission, and it was obvious to me that they believed their mission was especially important.

"At that time, every American serving as either Superintendent or Assistant Superintendent was a veteran. The weather was perfect and, as with my first visit 13 years before, the cemetery was in pristine condition. Additionally, that day there were thousands of visitors in the cemetery from all over the world. Many were D-Day Veterans who came back to Normandy to pay their respects to their fallen comrades. The Superintendent told me that visitors to the cemetery exceeded 1 million annually. I had grown up around a family of World War II Veterans and I really liked interacting with them.

"In the fall of 2000, I was the J-33 on the Joint Staff. It was a seven day-a-week job. One Saturday morning I was headed into work and ran into Colonel (Ret.) Ken Pond. He was leaving the Pentagon gym after his workout. I had known Colonel Pond when I was the XO and the Deputy Commander of the 4/325th in Vicenza, Italy. He was the Southern European Task Force Chief-of-Staff and the senior Infantryman on the Caserma Ederle installation. He was an immensely proud Airborne Ranger, had served two tours in Vietnam, and as an Infantry captain had served in the Mountain Ranger Camp.

"After he retired, he went to work for the American Battalion Monuments Commission, first as the Director of the Mediterranean Region in Rome, and later as the Executive Director and Chief Operating Officer in Arlington, Virginia. I told him about my visit to Normandy American Cemetery and I ended our conversation by saying, "Someday, I would like to work for the American Battle Monuments Commission."

"One day I was in my office in the Pentagon and my XO notified me that Colonel Pond was on the phone and wanted to speak to me. I picked up the phone and said, "Sir, how can I help you?"

"He immediately replied, "I think maybe I can help you!" He explained that the agency was looking to hire a new Director of the European Region in France. He went on to briefly explain a little about the ABMC mission and a little about the duties and responsibilities of the position.

He knew I was on the promotion list for major general, but he asked me to consider the job.

"As I said previously, I was extremely impressed with the cemetery in Normandy, and I really liked what he told me about ABMC's mission. I told him I would think about the offer and talk with Jennifer. I expected to be promoted to major general sometime the next summer, but after much discussion with Jennifer, we made the decision to retire. I have been asked many times why I decided to retire and join ABMC. Very simply, I really liked the mission of the agency which was to be the guardian of America's overseas commemorative cemeteries to honor the service, achievements and sacrifices of America's Armed Forces.

"I knew with my experience I could contribute to the mission, and I wanted to be a part of it. I was also very impressed with the employees, both the Americans and the host nation French I met who were working in Normandy. At that time, all the Americans working in the cemeteries worldwide were military veterans. The vast majority were Army Veterans, mostly former senior NCOs.

"ABMC wasn't exactly the Army, but it was run and staffed with military veterans, and I liked that a lot. The Secretary of ABMC was a distinguished retired Army officer, Major General John Herrling, an officer of impeccable character. I had complete trust and confidence in Colonel (Ret.) Pond, ABMC's # 2 man. I knew I would be comfortable working in that environment. As I later saw as I toured all the locations in the European and Mediterranean Regions, the staffs were extremely talented, completely committed, very focused, and totally dedicated to ABMC's very sacred mission.

"Also, in addition to the World War I and World War II cemeteries in France, Belgium, Luxembourg, Netherlands, and England, I would be responsible for the care and maintenance of some equally impressive memorials and monuments in Europe, to include the Pointe du Hoc Ranger Monument. I decided to join this winning team because their mission aligned with my values and world view."

Bill spent almost four years as the Director of ABMC's European Region and five years as the ABMC Executive Director and COO. It was a huge job. At the time, ABMC administered, operated, and maintained twenty-four permanent American military burial grounds, and twenty-four separate memorials, monuments, and markers, in seventeen foreign countries, the U.S. Commonwealth of the Northern Mariana Islands, and the British dependency of Gibraltar.

Every step of the way, he was supported by Jennifer, his spouse of over 50 years. Her story is shared in Chapter 15, Heroes at Home. Bill and Jennifer's daughter, Kristan, is married to an Airborne Ranger brigadier general, Todd Brown, who commanded 1st Ranger Battalion and the 75th Ranger Regiment. Their son, Pete, served in combat with each of the three Ranger Battalions. He took command of 2nd Ranger Battalion in June 2024.

Based on Bill's conversations with Todd and Pete, and his personal observations, I think it's appropriate to close this chapter with his thoughts on the intangible benefits of military service: "The first benefit is discipline. If you don't have discipline when you come into the military, you're going to get it. Regardless of what you hear about how the military is soft, this generation is soft, and this generation has gone woke, whatever that means, it's not true. When these kids come in the military, it's going to change them. At 18 years old, they are trying to find their way. They are going to learn where they are going and they're going to mature. They're going to find themselves in the military.

"I don't care what service they go into, but certainly in the Army, they will discover who they are. They're going to get this feeling of a family because the military becomes your family. So, if you didn't have a family, or you grew up in a broken family, you have one now. I continue to hear that from so many servicemen. You are certainly going to get smarter. You're going to learn how to push yourself and how to be part of a team.

"As I look at high performing organizations in the civilian world, who's succeeding and who's failing, the companies that are doing well have people who function as a team and are willing to subordinate their individual goals and ambitions for the good of the organization. Those are the type of things I think you learn in the military."

Jim Dubik and Bill Leszczynski are shining examples of the benefits of a full career of military service. While they were evolving over decades as consummate military professionals, they were expanding their world views and deep respect for others. This resulted in them living lives of purpose with continued selfless service to others in their noble civilian professions.

10. LEADING OTHERS

Leaders get paid to know their job, provide a clear "why" and resources to their soldiers, and to inspire and care for them daily. Be, Know, and Do!
—CPT Lawson Magruder

From left: Al Kovacik, Jim Smith

"ABSOLUTELY NOT! I TOLD YOU THERE IS NO WAY WE'RE DOING THAT," THE Sergeant Major said emphatically.

SSG Kovacik was standing at Parade Rest in front of the Sergeant Major's (SGM) desk. It was 1984. He was an intelligence analyst with the Pershing Nuclear Missile Battalion in Germany and currently attending the two-week, Primary Leadership Development Course (PLDC) for junior NCOs.

As he stood there in front of the SGM, he was gob smacked by the blatant racial animus that was on display. Something still infecting the Army since the Vietnam War. It was an awful cancer. As a junior NCO, he had little or no power. But he knew right from wrong and had incredible moral courage.

Al described the situation he found himself in. "When I went to

PLDC, there were a lot of racial issues, especially with the junior enlisted, but also with the school leadership. When I recommended that charges be brought up on a couple of other junior NCOs for participating in criminal activity, the SGM said no. I stood my ground. The SGM and one of his senior NCO instructors tried to intimidate me on several occasions during the course. I thought they were going to fail me out of school. But I didn't back down.

"I said to them, 'No Sergeant Major! I don't care if they're black, white, red, green, whatever. I don't care if you're wrong. This is not right and I'm not going to stand for it.' Well, long story short, the SGM pulled me aside near the end of the course and finally admitted I was right. I don't know why he changed his mind, maybe he did an investigation, I don't know? But they finally took appropriate action against those two junior NCOs. I wasn't trying to make a negative example. It was simply a matter of right versus wrong. I didn't care who they were, what they were, it's what they were doing that was wrong. I graduated and was awarded the Leadership Award."

This is what principle centered leadership in action looks like.

Alan D. Kovacik entered the Army in 1973. After high school, he had tried out college. He lasted a single semester. The Vietnam War was winding down and the college atmosphere was antithetical to the way he was raised. Anti-war cynicism was horrific, his grades were poor, and he wasn't enjoying himself. He realized college was not right for him now. His dad was a WWII Army veteran, so Al knew a little bit about the Army.

At the recruiters he saw a brochure with a cool-looking guy on the front. It was an Army Ranger. There he stood in green jungle fatigues wearing a patrol cap. The cap had a bright gold and black Ranger tab sewn across the front and a pair of Airborne wings sewn underneath. He was wearing a load bearing equipment (LBE) harness with magazine pouches and canteens attached to the pistol belt. A neatly coiled, long rappel rope was draped over his left shoulder, across his chest, its coils laying on his right hip like a bandoleer. He held an M16 in front of him, one hand on the pistol grip, the other on the hand guards, staring into the camera lens.

Al asked the recruiters what they did, and they said read the brochure! He did and asked, "Can I do that?" The recruiters checked and told him the Airborne Ranger Enlisted Option was available, provided he scored high enough on the entrance exam. So, he decided right there, he wanted to be an Army Airborne Ranger. He got his wish.

Al completed the Ranger pipeline: Basic Training, Advanced Infantry Training, Airborne and Ranger Schools by October 1973. The two Ranger Battalions had yet to be activated. Al and the other Ranger enlisted options graduates had only three choices for an Airborne Ranger assignment: Alpha, Bravo or Charlie Company Rangers, 75th Infantry. Al selected Bravo Company because they were stationed at Fort Lewis, WA and he'd never been there. After he arrived, he served for a little over a year, when seemingly out of the blue, the Army deactivated the Ranger companies and replaced them with the two, newly activated Ranger Battalions.

Ranger companies were small, with less than 100 Rangers. Battalions were six times as big. It sounded exciting. Once again, he had a choice, 1st Ranger Battalion at Hunter Army Airfield in Savannah, GA or 2nd Ranger Battalion at Ft. Lewis. He decided to stay at Ft. Lewis because of the hunting, fishing, mountain climbing, and camping.

He joined the 2nd Ranger Battalion in January 1975. He was assigned as a Squad Leader in 3d Platoon, Bravo Company. He was the 37th soldier to arrive at the Battalion, which eventually would be home to more than 600 Rangers. Al was assigned to lead the 1st Squad. By June, the company had most of its compliment of soldiers. Al led and drilled his soldiers into a solid team. By October of 1975, just two months before the Battalion's combat certification test or ARTEP, the platoon leadership decided to make some key personnel decisions, realigning some of the junior NCOs and senior specialists between the squads to strengthen the entire platoon.

Training began to intensify. With the help of his team leaders, Al picked up the training and continued to forge his squad over the next two months into a hardened fighting unit. His squad became one of the most technically and tactically proficient squads in the platoon, the company, and possibly the battalion.

Al spent two years in 1st Squad before being asked to take over as Platoon Sergeant of Weapons Platoon. The platoon had recently failed their Army Test and Evaluation Program (ARTEP), something that would not stand in the Company or Battalion. Nobody failed an ARTEP. It was a failure of leadership.

The Battalion Commander, LTC Wayne Downing, was not amused. The current company commander, CPT Jim Dubik, and the First Sergeant, SFC Roy Smith (formerly the 3rd Platoon, Platoon Sergeant), decided they needed to take action to get the platoon back on track. They knew they needed an exceptionally strong and experienced leader. They immediately thought of Al. Al was one of the strongest squad leaders in the Battalion, he was also technically and tactically at the top of his game. As an exemplary leader, he was virtually unchallenged. He agreed to take Weapons Platoon without hesitation.

In his own words, Al said, "So I assumed the responsibility of Platoon Sergeant. Within three or four months, we turned the platoon around and maxed the ARTEP. The guys knew what had to be done. They were very knowledgeable, motivated, and professional. They just needed some guidance and steady, reliable leadership. We had a great time. During that time, we had the 90mm recoilless rifle and no one knew how to engage targets at night because all we had was an illuminated reticle. Well, we came up with an idea and we worked it out. We were the first Weapons Platoon in the entire Battalion to demonstrate how to engage targets at night with this new technique that we devised. Lieutenant Colonel Downing was just ecstatic, and CPT Dubik couldn't have been prouder of us."

Humility, listening to your men, and understanding their strengths and weaknesses are the key ingredients of exceptional leadership.

Al came down on orders to be a basic training Drill Instructor (DI) at Fort Benning, GA. Unknown to Al at the time, his military career was about to drastically change. After being an Airborne Ranger for almost five years, he was off to the regular Army. Something most of us dreaded, but new was a fait accompli. After three years on "the trail," Al ended up getting his next assignment to guard the Persian II Weapons System, the Army's primary nuclear-capable theater-level weapon, in Greece.

Disappointed isn't a strong enough word. Al tried his hardest to get out of it. He requested to go back to one of the Ranger Battalions, to the Ranger Department, or to Special Forces. After training recruits for three years, he wanted to get back to his Airborne Ranger roots. As with many things that occur in the Army, they had a slot to fill, and he drew the short straw.

Al made the hard decision to get out, which meant he would have to sign a waiver saying if he ever decided to come back in the Army, he would lose two grades. So, he turned down Sergeant First Class (E7) and left the Army as a Staff Sergeant (E6). Turning away from his dream to continue serving his country as a combat warrior was not easy. After about a year in the "civilian world," Al wanted to be back with his beloved Army.

He joined the reserves where he could keep the rank of Staff Sergeant and then accepted an active-duty Drill Instructor (DI) position at Fort Leonard Wood, Missouri. It was good to be back. He ran into DI friends he had served with at Ft. Benning. After talking to them about coming back into the regular Army, Al decided to re-enlist.

In his words, "They offered me three choices. Combat Engineer, Intelligence Analyst, or a Pershing Missileman. I didn't want to do Pershing because that's the reason why I got out in the first place. I didn't want to be an engineer because I would stay at Ft. Leonard Wood for training. So, I asked, 'Out of the three, which one gets promoted fastest?' They said, 'Military Intelligence.' I said done. I was off to Fort Huachuca, AZ."

After Al was there a few days, he realized there were some serious leadership issues in the company. He approached his First Sergeant, who was an excellent, no nonsense, senior NCO. I said, "First Sergeant, there's things going on in the company and battalion that you're not aware of. Look, I'm either going to talk to you tonight or I'm going to talk to the Inspector General in the morning. He said, 'Whoa. Okay. I'll tell you what, I'm going to be back in my office in 15 minutes. You come and see me.'"

Long story short, Al explained to him what was going on in the organization. Al went to work the next day and by the time he came back to the barracks, numerous NCOs and a couple junior officers from their sister companies had been relieved, with some being arrested for fraternization. The Battalion Commander and Al ended up talking about Al's leadership and moral courage for coming forward. He learned about Al's experience. The Commander was so impressed, he offered him a job as a school Cadre NCO once he completed his AIT training.

Eventually, Al came down on orders for Germany. His assignment, the Pershing II Nuclear Missile Battalion. This is where our story started. Al, a former Airborne Ranger Staff Sergeant and Drill Instructor, promoted once again to Staff Sergeant (E6), had to confront a Sergeant First Class (E7) and a Sergeant Major (E9) about leadership. This is the definition of what leaders are supposed to do.

About three years later, Al left Germany and was assigned to the 1st Cavalry Division at Ft. Hood, TX. In late 1990, the division deployed to the Kingdom of Saudi Arabia for Operation Desert Storm. Following his tour there, he was reassigned to Alaska to help stand up the 6th Military Intelligence (MI) Battalion for the 6th Infantry Division and decided to retire in Alaska in 1996 after serving as both an MI First Sergeant and the G2 Divisional Intelligence, Sergeant Major.

In retirement, Al went on to become a contractor supporting the Army Maneuver Training Center, training soldiers on a variety of computers used for different combat systems. He officially retired from his civilian job in 2018 and now spends his time in the great Alaska outdoors with the second love of his life: hiking, camping, hunting, and fishing. He reserves his first love for his wife Kim and his two grown children, one a former

Special Forces NCO who owns a consulting company and his daughter and their nine grandchildren.

Like so many of the soldiers that started in Bravo Company, Al's story is a textbook example of the Abrams Vision—send Ranger NCOs and officers to the conventional Army and as leaders imbue those Ranger values and leadership attributes and characteristics to rebuild the force. Simply put, Al took the values and standards from the 2nd Ranger Battalion, and Bravo Company, and applied those in the most unlikely organizations.

In his case, he went from being part of the 'tip of the combat spear' in the Rangers, to support units that were more technically focused organizations. In every case, Al successfully led combat soldiers and technicians to be coherent, highly motivated, functioning teams. Al built teams like this in every organization in which he served over his military and civilian career spanning 36-years. All resulting in more competent, more professional, and mission focused units. Al is the living embodiment of the Abrams vision.

When asked how his time in the 2nd Ranger Battalion effected his leadership style for the rest of his time in the Army into his civilian career, Al said, "When the quarterback is going to throw that ball on a down and out to a certain player and that player is the best one, you know he's going to make that catch. We were always that player. I might have taken that for granted after a while and expected it because we never had to talk about it among us. But that wasn't the way it was in big Army, so I had to build it."

Leadership is required in every organization in the world. Many in the civilian world call leadership 'management.' But they are wrong. Management involves a focus on executing functions to accomplish a strategic vision for the organization. Whereas "leadership" is the art of inspiring others to willingly follow you.

***

Jim Smith is another example in the art of leadership. Jim Smith got a call to have lunch with the Oregon Trail Council President. Over lunch, they discussed the challenges facing the Council.

"Jim, you know we have some serious challenges ahead of us. Between all the abuse scandals over the years, and most recently COVID, our Scout Master and youth participation is tanking. I need your help."

"What do you need me to do, Randy?" Jim replied.

"Our enrollment is struggling given all the issues surrounding the lawsuits and Chapter 11 resolution. Now that we've implemented new training policies for our Direct Contact Leaders, our non-Contact Leaders, and the Youth Protection Training, we need to get the news out to the community about the positive changes we're implementing in the Boy Scouts. I know you're busy with your school board, the District Council and your District Units meeting the new Oregon Trail Council requirements, and of course it's rechartering time..." Randy continued.

Jim interrupted, "Randy, I already have rechartered 14 out of my 15 units and we've been focused on getting the Scout Masters fully trained all year. While it is challenging, we're making really good progress. The bottom line is that even with just half my normal team, my District is leading the entire Council in all areas!"

"I know, that's why I called. I also know you've been heavily involved with your school board for years, and now as Chairman, you're leading the process of the staff and community to turn around your failing school district. We need your leadership now to be the Council's Liaison to the different State School Districts to tell our story and show how serious we are about rebuilding membership. Parents are concerned, so we need a solid plan and you're the guy to tell them that they don't need to worry about their kid's safety. That we have trained Leaders at all levels. We really need your leadership on this Jim, can you help us?"

"You bet, Randy. It would be my honor," Jim replied.

Jim summed up his challenges to turn around a failing Scouting program while having to push back against some Council Leadership who wanted to reduce standards, "Well, the co-Chair said 75% goal to train adult leaders was good enough. I thought to myself, what would Eldon Bargewell do? I simply said 75% wasn't my style, so I ruffled feathers and the complaints piled in that I was being pushy. I let the executive committee know that if 75% was the new Council standard, I would resign. Rob, our acting Scout Chief Executive, and the other executives, agreed 75% was unacceptable. He said to me, 'Good, but remember, you can't leave before me, you can ONLY leave after I leave, but not before.' It was a nice compliment."

To say Jim has played a significant role in his Boy Scouts Council leadership, bringing positive changes to units training levels and ensuring their success during the last few years of challenging times, is an understatement. He did this while also serving as the Chair of his School Board during COVID 19. His dedication and ability to inspire others have made him an irreplaceable and influential figure within the community, his local education system, and state scouting communities.

So, how did he accumulate the knowledge, skills, and abilities to become an indispensable community leader?

James Ray Smith's Airborne Ranger journey started in the Boy Scouts. Jim described himself at the time as being "a 19-year-old, rambunctious, over motivated, and sometimes scout." He was influenced to join the Army by his Scout Master, John, who was a Reserve Army Command Sergeant Major and had been an Army Air Corps B17 crewmember in the Pacific during WWII. John had a lot of influence over Jim and encouraged Jim to join the Army and try out for the Rangers before he got into too much trouble.

Jim enlisted in the Army at age 19 in May 1975 from Roseburg, OR. Unfortunately for Jim, his recruiter falsely told him there were no direct unit assignments to the Rangers, but he could enlist for Station of Choice. Jim chose Ft. Lewis, WA because he was told the 2nd Ranger Battalion was stationed there. Plus, he got to stay in his beloved Pacific Northwest. After he successfully completed Basic and Advanced Infantry Training, he arrived at Ft. Lewis in August 1975 as a "mosquito wing," (Private/E2) leg (non-Airborne) infantryman.

Upon arrival at the 525th Replacement Depot, he saw a long line of about 100 soldiers, mainly Privates (E1-E3) and some Specialists (E4), standing along a wall in a hallway leading up to an open office door where a clerk was sitting in the room. He found out that they had been assigned to the 2nd Ranger Battalion and had decided to voluntarily terminate their jump status. Apparently, jumping out of perfectly good airplanes with 100 pounds of light-weight combat gear while wearing a 50-pound parachute in the pitch back darkness wasn't their idea of having fun. Neither were the weekly, 20+ kilometer forced marches and 10 mile runs. Jim was determined that if he got to the Rangers, he would not be one of them.

A few days later, Jim successfully interviewed with the 2nd Ranger Battalion Command Sergeant Major and was accepted. In early September he was assigned to 2nd Squad, 3d Platoon, Bravo Company—just another non-airborne qualified private being thrown into the grinder. But nobody who arrived as a leg stayed a leg for very long. About a month later, Jim and about 30 other "leg," want-a-be Rangers, were off to Airborne School for their first test. Jim returned in early November as a fully qualified paratrooper. It was a piece of cake.

Just a few short days after he returned, he was packed, inspected, and ready to make his "cherry jump"—his sixth jump since finishing jump school less than a week before. This jump would be his right of passage. He would be a real paratrooper. It was called a "mass tactical" jump. That meant there would be multiple C130 Hercules Cargo Aircraft flying in formation. Each aircraft was slightly offset from the other with about a minute separation between them. The C-130 could hold 64 combat equipped jumpers who were crowded shoulder to shoulder sitting on the narrow, nylon webbed benches that lined the outside and middle of the aircraft. To say it is tight is an understatement.

They were parachuting into the fictional country of "Pineland." The mission was to test each platoon in Bravo Company and the Company's command and control. This was to get ready for their combat certification Army Readiness Training and Evaluation Program (ARTEP) coming up in December. It would be a combination of light infantry, squad, platoon, and company size attacks, company defensive operations, and small unit Ranger patrolling operations like raids and ambushes.

The grueling training event would move the company through rugged terrain, thick Pacific Northwest rain forests, dozens of miles of patrolling, forced marches, assaults, defenses, raids, and ambushes. The weather forecast was grim—it would be prophetic. In October-November the weather turns schizophrenic in the Pacific Northwest. It can go from warm sunshine, to pouring rain, sleet, and snow and freezing temperatures in a matter of a few hours and sometimes minutes. By the end of the week, it left most of the Rangers dangerously close to hypothermia.

The night drop would be from 1,000' above the ground. It would be a moonless night. Each jumper donned many pounds of gear hours before the jump. Sitting on the cold, damp tarmac waiting to load the plane, the older hands of the company joked, slept, ate, smoked and steeled themselves for what they knew would be an arduous week. Their attitude, "If you don't mind, it don't matter"—that's the mental game. It was all a mental game. Jim was about to find out if he had what it took or would he be one of those guys standing in line looking for a new assignment "down the road in leg land." He was also about to learn that jumping was the easy part of the job. When the time came to jump, Jim and 63 of his closest friends jumped into the black of night, emptying the aircraft in less than a minute.

Over the next six days there was little time to sleep, rations were eaten cold, the ground was nearly frozen so digging armpit deep fighting positions with miniature shovels called "E-tools" was exhausting. They'd get them done, then it was time to move out. "Fill your holes in, ruck up, we're moving out!"

After six days and nights of no sleep, marching and running mile upon mile, the exercise culminated in a company size attack on an enemy defensive position located on a ridge. The direct assault required that the main assault force had to low crawl, online, across a 500-meter open field full of "cow paddies," without being detected.

Exhausted Rangers were falling asleep on the ground as they crawled, numb to the bone from being wet and cold. Their feet were blistered from the miles they had walked for the past six days, and they were looking forward to the ride home to warm barracks and hot chow. Just another day in the Rangers. "If you don't mind, it don't matter. It ain't nothing but a thing."

Jim thrived and rose quickly in the platoon. Growing up in Oregon, Jim described his upbringing, "I grew up in Oregon doing stupid things as a boy, like playing with a beach ball in the Cascades, in a mile-high glacier pool." Rangers learned quickly that the first thing a leader had to learn to endure were hardships, hardships made you resilient and humbled. Hardships grew confidence, helped manage fear, and made you listen and learn from those more experienced than you.

Jim turned out to be a natural student. He started off as a rifleman, a grenadier, became a team leader, and then a squad leader.

In Jim's words, "I was proud that it was a tough place to be. And I remember mornings, rolling out to the PT formation hoping it was going to be an easy day, only to find out it was going to be a ten-mile run this morning and I might have had a little too much to drink the night before. And I look back on that, I had no problem physically making it and doing what was necessary to be there, but it was every day—every day you had to give 100%.

"So, you develop a mental toughness along with physical toughness. But it was about the leaders we had. I describe it as an opportunity to walk with giants. There were giant leaders in the battalion at that time. And their job was to make us the best leaders we could be. I remember my Squad Leader, SSG Jimmy Bynum. He would take the privates and specialists and do 'Squad Leader for the Day' training. Putting one of us in his position to lead us from wakeup until bed. He would critique us that evening about how we did and what we might consider doing differently. He mentored us, set us up for success, and we all became successful.

"I took that with me into the civilian world. Not long ago, I was the School District's Board Chair. It's a two-year term with a two-term limit and I was ending my final term. They extended me for a few months due to a district emergency. Another board member asked me what my plan for succession was since I had reached my limit. I said, my plan is you're going to become the Board Chairwoman and I'm going to mentor you. So, I stepped down to be the Vice Chair and she was appointed to the Board Chair position until the next election. She voiced concerns and I said, you'll do fine. You're smart, you're capable. And she was."

Jim spent three years in the Battalion. He could have gone into Special Forces but chose the harder leadership path. He said, "I stayed in primarily because there were other places I wanted to go in the Army and other things I wanted to do. One of the most important things that resonated with me were leaders I respected who told me, when you go out to the regular Army, that Ranger Tab is going to get really heavy. They are going to lean on you a lot and ask you to do things above your rank. And I found that to be true. When I look back at my experience and think of General Abrams' vision about using Rangers to rebuild the Army, I realize that was my experience. That's what we did. That's what I did.

"Those years in Battalion gave me a great sense of confidence. When I left the Battalion, I was assigned to B Company, 2nd Platoon, 1st Battalion, 35th Infantry in the 25th Infantry Division, in Hawaii, as a Platoon Sergeant. After about 60 days, the First Sergeant came to me and said he needed me to take over 3rd Platoon as their Platoon Sergeant. He also told me I was going to be the acting Platoon Leader since they were losing him as well. I asked him when the platoon leader's replacement would be coming in and he said he didn't know. I would do both jobs until further notice. I did both jobs for the next two years."

Jim told us one of his favorite stories of when he was in the 25th, "Here's a funny story. There is a picture of my first platoon leader, Eldon Bargewell, shaking hands with General Schwarzkopf in Kuwait in 1990. He was a First Lieutenant when we served together in the Battalion. In the picture, he is a Lieutenant Colonel. For me that picture represents the full circle of the training I received during my time in the Rangers.

"What Eldon didn't know of course, is that I and two other former Ranger NCOs trained three platoons from the 25th that were providing personal protection for General Schwarzkopf. Unfortunately, just before deployment I had been reassigned to Division Headquarters as the G3, Operations Tasking NCO and didn't get to go to Desert Storm with them. Otherwise, I would have been able to reconnect with Bargewell."

Jim went to serve multiple tours in Hawaii. In between tours in Hawaii, he went to Ft. Benning, GA as a Ranger Instructor. Jim would serve with many of his fellow 2nd Battalion Rangers over the years, both in the Ranger Training Brigade and other assignments. Jim retired in 1996 after 21 years, culminating his successful career at the University of Oregon's Reserve Officer Training Corps (ROTC) as the Department Sergeant Major/Senior Instructor and Ranger Challenge Team coach.

He ended his military career like he began, providing all he led with a solid NCO leader applying his well-honed leadership and training skills to make them technically and tactically proficient. The two most consistent themes in Jim's adult life, during and after the military, are leadership and training to standards.

After retiring, Jim started a business as a pet store supplier and exotic bird breeder, which he ran successfully and profitably for 14 years. He had amassed nearly a quarter million-dollar collection of exotic birds. It all came to a sad end after the economic meltdown of 2008. Many of his customers had to close their doors because they couldn't get bank loans and went out of business. As he lost customers and revenue, he had no choice but to close his own doors as well. Jim had decided to focus on helping his community.

A new leadership opportunity was upon him. Shortly after he had closed his business, he was asked to be on the school board to help pass a school bond to build a new school. He said of that time, "The biggest influence on me was the day that President Bush stood up after 9-11 and said, 'get involved, do something'. Uniform service, become a fireman, a policeman, et cetera. And a lot of the guys I knew from the Army were deploying to Afghanistan and working as civilian contractors. I had adopted two boys over the years from fostering kids, which I did for over 16 years, 34 kids in total. My new wife and I had also had two young children, so when I suggested going overseas, she killed that idea.

"So, I decided to refocus and take the offer to interview for an upcoming board vacancy. I got involved in the budget committee for the school district to understand their financials. When the seat became available, three of us interviewed for the position of 'member at large' until the election. I was selected. When the election came, I ran for it and won. I've been on the school board for 22 years and was recently asked by our superintendent to run again in 2025 to finish an educational facility project I had led for the past decade and to keep the team together."

There's so much more we could write about Al and Jim, but it would take a book of its own just about them. They were front line warriors to help rebuild and transform a demoralized Army. Al went on after retirement from the Army providing critical training for our soldiers going to combat in Afghanistan and Iraq for the next 17 years.

Jim transformed his small community, got a new school built, created another large agriculture training facility for the district, and continues to lead his State's Boy Scout Council District out of the wilderness of shame, setting and enforcing new standards, and bringing them into an exciting new era in the 21st Century.

This is how you turn young men and women into compassionate leaders. Shared hardships, training to standards, demanding expectations, patience, mentoring, leading by example, and time—it takes time to meld, forge and temper leaders.

One of the first lessons Rangers learn is to physically and mentally accept hardships. Learning just how invaluable it is to learn adaptability and resiliency. Learning how to endure hardships helps identify your weaknesses, which in turn allows you to face and work on them constantly.

Then you make small improvements through dedication, discipline, and perseverance. The upside—it builds self-esteem, which builds belief in yourself, which builds self-confidence, which builds success.

Suffering is the warrior's tool to develop and grow.

Al and Jim are emblematic symbols of leadership who represent the best in all of us and the U.S. Army NCO Corps.

"Competence is my watchword. My two basic responsibilities will always be uppermost in my mind—accomplishment of my mission and the welfare of my Soldiers. I will strive to remain technically and tactically proficient. I am aware of my role as a noncommissioned officer. I will fulfill my responsibilities inherent in that role. All Soldiers are entitled to outstanding leadership; I will provide that leadership. I know my Soldiers and I will always place their needs above my own. I will communicate consistently with my Soldiers and never leave them uninformed. I will be fair and impartial when recommending both rewards and punishment."

—From the Second Stanza of the U.S. Army NCO Creed.

11. SHARED VALUES: TRUST & TEAMWORK

It is the simple things that build trust: doing what you said you would do in a timely manner; not allowing double standards; treating others in a demanding but not demeaning manner; and admitting mistakes. Once trust is established as the foundation, incredible teamwork will follow.
—CPT Lawson Magruder

From upper left: Jim McNeme, Tim Grzelka, Jim Jackson

ANACORTES, WA—SKAGIT VALLEY HERALD, APRIL 20, 1977

The Texaco oil refinery on March Point was successfully demolished this past weekend. A group of elite Army Rangers successfully attacked and sabotaged the refinery early Sunday morning. The exercise was part of a special training program for the 100-man Airborne Ranger unit at

Ft. Lewis. "The Rangers train to become a fast-action strike force, ready to move anywhere at any time," said Sgt Jim McNeme.

Refinery Manager Phil Templeton said, "I don't know how they got in and out without anybody seeing them. The Texaco security force was alerted to the assault ahead of time. Yet somehow the troops climbed over the surrounding fences and were able to place dummy explosives on refinery equipment such as boiler facilities, pipelines, railroad lines, and other strategic points."

Trust and teamwork are not something that happens overnight. It has two elements: trustworthy, competent, and steady leaders, both officer and NCO, and well trained and disciplined troops. These cannot be imitated. These are elements set from the very top of a unit and passed down the line to the most junior privates. They are unit wide values, standards, and expectations. And it is months, if not years, of individual and team training. It involves hundreds of detailed mission rehearsals, for dozens of different operations. Training in all types of terrain and year-round weather conditions both day and night with firmly established Standard Operating Procedures (SOP) at all levels of operations.

It is thousands of hours of doing those things over, and over, and over again until it allows you to perform without speaking because you already know what each other is thinking. This is how one hundred men sneak into and out of a well secured and massive oil refinery without being seen or heard—Trust and Teamwork.

We will highlight three Rangers that are sterling examples of trust and teamwork, Jim McNeme, Tim Grzelka, and Jim Jackson.

James McNeme grew up in Dallas, TX in a patriotic and loving family. His father was a WWII veteran. One thing was made very clear during his upbringing—duty was a big deal in the McNeme family. Jim graduated high school in 1961 just as the Vietnam War was beginning to heat up. Following in his father's footsteps, he sought a degree in civil engineering. During his sophomore year in college, he got married. After college Jim, and his wife Patti, went to Houston, joined a large engineering firm as Engineering Manager, and stayed there for four years. He loved his job.

The desire to serve in the military was extraordinarily strong in Jim and Patti, as it was in their community, and the church they attended. So, at the ancient age of thirty-one Jim and Patti made a hard decision to take a leave of absence and do their part for their country by joining the Army.

The Army offered Jim a commission in the Corps of Engineers, but he declined. He wanted to be a Ranger. He enlisted in the Army in March 1974 for the 2nd Ranger Battalion. He arrived at Bravo Company in February 1975 after completing Basic, AIT, and Airborne School as an E4. He was assigned to 1st Platoon and was one of only a handful of enlisted college graduates to join the Rangers.

He attended Ranger School in November 1975 and went back to 1st Platoon as an E5 team leader. In late fall 1976, he was reassigned to Headquarters Company as the Training NCO. As Jim tells his story, "It was a life-changing event. It was my niche. As a Civil Engineer it was all about details and as the Training NCO, it was all about details."

The title Training NCO does not describe the responsibilities of that one-man job. As part of his duties, Jim was responsible for managing and scheduling mandatory enlisted professional development schools, maintaining the company's advanced individual training Order of Merit List (OML) for coveted skill qualification courses like Ranger School, Pathfinder School, Combat Diver School, Military Freefall School, Sniper School, and other advanced training for both enlisted and NCOs.

However, the thing he loved the most was participating in the operational planning for the company. He believed that all training missions needed to be realistic and required everyone to think. And that brings us back to the beginning of this chapter. Several months before the training mission, the Company Commander had told Jim what he wanted to do—"I want to raid an oil refinery. Make it happen."

This was a BIG deal. Jim began to plan the various elements to meet the commander's training objective. It became an immensely complicated operation with many moving parts, primarily because of the location of the target—a busy peninsula in the Puget Sound of Washington State.

It is important to understand the level of trust and confidence the commander had to have in Jim as a junior NCO. It is significant. As a rule, commanders must delegate to make things happen. In doing so, if something goes awry, they alone remain accountable. This is where trust, open communications, and teamwork all play a part. Trust is only earned over time through demonstrating trustworthiness. This is both from the commander down and the subordinate up.

There was no doubt the commander trusted Jim. He had demonstrated that he had the knowledge, skills, and abilities to organize an epic training event. Jim's years of planning and executing multi-million-dollar civilian engineering projects, both during his college years, and for a decade before joining the Army, had prepared him for a large, complex project like this. Once he secured a target, then he could sit down with the commander and finalize the commander's intent for the training event. And that is exactly what happened.

It is important to understand this took months. But in the end, co-ordinating the use of one of the largest west coast oil refineries on the Washington coast as a target was straight forward. The facility manager was 100% onboard. Now that Jim had identified an acceptable target, the command element could make an initial concept plan to begin arranging the other logistical elements of the mission. The refinery's location on the peninsula allowed only for a seaborn approach because the land approach-es were all blocked by large industrial parks, highways, and other highly trafficked urban areas. The only tactically sound way to clandestinely get to the target was to come from the sea in rubber boats.

This required backwards planning. Starting from the target location and working backwards to the Ranger facility at Ft. Lewis. There were levels of complication that required coordination and timing. How large of a raiding force? How big of a seaworthy vessel was necessary to transport the raiding force and their rubber boats? Where to drop off the force at sea that was far enough away so to avoid being detected from the target, but close enough so they didn't have to paddle all night?

How to get the raiding force from Ft. Lewis to catch their ride to the drop off point? Jim put all the elements together, coordinating helicopter transport from local aviation units to fly the 100+ man raiding force and their rubber boats from Ft. Lewis to the U.S. Coast Guard Station at Whidby Island. Arranging a Coast Guard Cutter to transport the raiding force to and from the ocean drop off site three hours away. Making sure the challenge of getting in and out of the oil refinery met Ranger standards by having a competent and alert opposition force.

Days before the actual mission, Jim arranged the training areas for day/ night rehearsals, building realistic mockups, and arranging transportation to and from the training areas. Rehearsals were key to hashing out the operational and tactical minutia to eliminate as much uncertainty as possible. All came together that cold, wet, April night, leading to a flawlessly planned and executed mission.

Jim would return to his civil engineering career in Houston and for the next four decades travel the world over planning and managing multimillion dollar projects. He is now retired and focusing his time on his grandkids, his church, and his community. In his understated way, Jim simply described his time with the Rangers, "I loved the Hooah stuff. At 80 years old, I still think I'm Rambo! I loved being a Ranger."

Trust comes in many forms. It is trust that builds teamwork. Teamwork builds confidence. Confidence builds competence. Competence results in mission success. The following story will demonstrate confident and competent teamwork, even under circumstances when it was doubted by leadership.

* * *

It was so dark on this December night you could not see your hand in front of your face. A schizophrenic storm had been raging off and on for days. The relentless freezing rain, sleet, and snow driven by fierce winds moved the giant conifer trees of the forest like they were tiny saplings. The temperature was near freezing, if not below. The Rangers were moving tactically, in a long single file, towards their final objective. The mission was to conduct a deliberate, battalion size infantry attack against a well dug in mechanized infantry opposition force. This was the final mission, of the final night, of the battalion's first, weeks long Combat Certification Evaluation, which would make the unit deployable to fight for America in any corner of the world.

The weather was the worst enemy. The men were soaked from weeks of relentless rain, sleet, and snow, river crossings and rainforest swamps. These were the days before high tech rain and winter weather gear like Gortex. The Rangers wore only their camouflage jungle fatigues, cotton field jackets, and jungle boots. These were hard men, but hardness did not matter. The lack of adequate cold and wet weather gear had its effect causing everyone to suffer from hyperthermia and frost nip. In some cases, it required a Ranger to be medically evacuated.

1st Squad, 1st Platoon, Bravo Company led the assault elements of the battalion on a winding, 4+ hour long movement to the objective. They were to be in their attack positions NLT 0300, with the attack commencing at 0400 on 15 December 1975. The terrain was tortuous, with deadfall (fallen trees) littering their path—many taller than a man.

The column was on time, despite the weather. Then they ran into this minefield of giant conifer toothpicks strewn across the forest floor. It made it almost impossible to follow a straight compass azimuth for more than a dozen meters. These were the days before GPS and the Rangers had to be dead on accurate to ensure they arrived at the exact point of departure for the attack. Traveling at night in pitch black, the terrain is invisible, so it was impossible to navigate around them. You just had to go through them. Navigating in such conditions required dead reckoning along a fixed compass azimuth. It was not hard to stray left or right while maneuvering up, over, under or around these giant fallen trees, which could make the column tens or even hundreds of meters off course.

1st Squad Leader SGT Tim Grzelka and his Alpha Team Leader, SGT Frank Rekasis were on point leading the 600-man attacking force to its departure position for the final assault. Both Tim and Frank were veterans of Company B Rangers, Long Range Reconnaissance Patrol (LRRP), a relic of the Vietnam War. Old Company B NCOs and officers helped form the core of 2nd Battalion, along with other veterans of the Vietnam War. These leaders taught their young proteges the right way to do things, including how to navigate with a map and compass. And then they practiced and practiced in all kinds of weather and terrain until they perfected their skills.

As Tim explained his experience on that dark and difficult night, "SGT Frank Rekasis and I were on point. We knew the azimuth and Frank said to me, 'You good?' I said yes. During the movement, our Platoon Leader, 1st Lt. Joe Wishcamper frequently questioned us, asking, 'Sergeant G, I think you're going off track.' We felt he wanted to take over for us. This went on for a while. When he came up again, I said, 'Sir, with all due respect you put us on point, we are going to take you right to the departure point. Just leave us alone. Let us do our thing. Frank and I are a team.'"

The column finally exited into a less dense area of the forest, so they could move out a bit faster to get there on time. As noted, time and accuracy were imperative. If they were off course, the mission could fail. If the mission failed, the battalion could fail it's combat certification. There was a lot at stake. If the battalion failed its certification because the guy navigating screwed up, it could end careers for leaders. It was that serious.

The Rangers continued to move and arrived ahead of schedule. Tim commented, "We nailed it! We arrived exactly where we were supposed to be on time. And that really set a lot of emotion in mine and Frank's heart, but especially me. We accomplished the mission right there!"

Lawson Magruder, the primary author of this memoir, was Bravo Company Commander. He reflected to Tim, "I had dinner with Joe at our last reunion and he brought up about his faith in his NCOs and challenging them during the ARTEP. He was looking over your shoulder and you all just basically said, 'HEY boss, we're on it.'"

Tim replied, "Yes! The first couple of times he wanted to use his authority and take over. And finally, I said, 'SIR! We got this. Just let us do our job.' And he did. He did not bother us again and we nailed it right to the wall." Lawson jokingly replied, "He probably had the company commander saying, hey, check your man out on this."

Everyone was earning trust that night. Lawson was responsible for getting THE entire battalion to the objective. Joe's 1st Platoon had the mission to navigate. Tim's squad was picked to be the tip of the spear. Failure was not an option, and it took enormous trust in these two young NCOs from both officers. They had to trust in their subordinate's knowledge, capabilities, and experience, which is hard sometimes when the stakes are so high. This is why trust and teamwork are so important and the best way to build it is to train the way you are going to fight.

Timothy Michael Grzelka came from a military family. He said what motivated him to join was his father, uncles, and other family members who served in WWII. "My father was a U.S. Marine in the Pacific theater during World War II. He served in the battles of Guam and Iwo Jima. We were talking about the service. He said to me, 'You know what? Service is a good thing. But when you jump off that landing craft and hit that beach, you don't know what's going to happen.' I thought about that and read articles about what it was like being Airborne, Special Forces, and jumping into combat. I thought the chances of survival would be a lot better jumping in rather than running down a ramp like my dad. I wanted to serve my country, and after talking to the recruiter, I chose the Ranger Enlisted Option."

Tim enlisted on February 20, 1973, on a three-year contract. He completed Basic, AIT, Airborne School, and Ranger School successfully and in November 1973 was assigned to Company B, 75th Rangers, Long Range Reconnaissance Patrol Company (LRRP) at Fort Lewis, WA. This is where Tim fully began to understand the importance of trust and team-work. Their patrols were small teams of 6-8 men. They would clandestinely infiltrate behind enemy lines and collect intelligence about enemy forces, and when needed, conduct ambushes, or call in tactical air support and ar-tillery on the enemy. During Vietnam, LRRPs were often suicide missions since the small teams could only carry so much fire power. This is why trust and teamwork were as imperative as technical and tactical expertise.

In October 1974, the LRRP companies learned they were being dis-banded and that the Ranger Battalions were being stood up. If they want-ed to stay in the Rangers, they would have to join either 1st or 2nd Ranger Battalion. They were conducting interviews for anyone in the company who thought they might like to join. Tim interviewed for the 2nd Battal-ion, and he was selected. Tim was assigned to 1st Platoon B Company, LT Joe Wishcamper's platoon as Squad Leader of 1st Squad, bringing us back to the beginning of his story.

It was a transformative time for the Army, which was transforming from a conscription force to a professional force. The Ranger's other mis-sion was to become the standard bearers of this new professional Army. To accomplish this, training was transformed by establishing exacting stan-dards and training to those standards at the individual, squad, company, and battalion levels. This model also built the trust between teammates and units which continues to this day in the modern-day Rangers, 50 years later.

In 1976, Tim decided to leave the Army. He pursued a career in the medical field. Shortly after getting out, he worked in a dialysis clinic as a technician. He learned that Penn State University, just 20 minutes from his home near Pittsburgh, had a biomedical engineering program. Using his GI Bill, Tim went into the program while working full-time at the clinic for the next five years. When Tim's clinic director learned what Tim was doing, he sat him down and asked him what his plans were. When he learned Tim's background and goals, the director of the clinic made him the supervisor of all the other technicians at the clinic, even those more senior.

It was about how Tim carried himself, his competency, his attitude as a team player, and the fact that he was reliable, dependable, respectful, and confident. All the traits forged in him during his time in the Rangers. They were an integral part of his character then and still are today. These are the simple secret ingredients in the recipe for success. It made him stand out among his peers. The director knew he could trust Tim at his word.

He graduated after five years as a biomedical engineer and joined a company repairing and maintaining hospital equipment. He first started with kidney dialysis and X-ray machines. Eventually, Tim decided to specialize in heart pumps and would spend the next 40 years helping to save people's lives all over the country by installing, repairing, and maintaining the equipment. The doctors so trusted him, they often would call him in the middle of the night for help to troubleshoot a problem with one of his lifesaving machines. Tim credits his time in the Rangers for his success. He said of his time there, "The Ranger battalion instilled in me, to this very day, the purpose, drive, commitment, and desire to accomplish the mission. Once you have those, and a positive attitude, you can do anything you want."

Today Tim is retired and lives with his wife in Bozeman, Montana. He competes in long distance rifle competitions and teaches young folks the art of long-range shooting.

* * *

The first time Captain Lawson Magruder observed Lieutenant Jim Jackson was in 1973 when they were assigned to the 2nd "Falcon" Brigade of the 82nd Airborne Division. Lawson was Jim's evaluator for the tactical maneuver phase of the brigade's annual best rifle platoon competition. Jim's platoon had been selected out of nine rifle platoons in the 3-325th Parachute Infantry Battalion to compete against the best platoons from the other two battalions. For six hours, Lawson observed Jim conduct detailed planning and preparation, and skillful maneuver of his forty paratroopers in a daylight attack. He was impressed with Jim's calm, confident leadership style and his ability to inspire his soldiers. He also marveled at how a leader who was 6 feet 6 inches tall could be so agile in the field.

It was clear that the platoon's exceptional teamwork was the result of the trust Jim had built within the unit. Little did Lawson and Jim know that 18 months later they would serve together in Bravo Company 2nd Rangers.

James T. Jackson was an "army brat" growing up as the son of a career soldier. His dad served for 33 years as an armor officer during World War II and the Korean War, retiring as a full colonel. He never pushed Jim to join the Army, but his example and the military lifestyle inspired Jim to join Army ROTC at Kent State University where he excelled in his military studies while also playing on the basketball team his freshman year. It was in both the ROTC and athletic worlds where Jim was imbued with the importance of building trust that will lead to winning teamwork.

He was commissioned an infantry officer in 1971 and began his 32-year journey in the Army. After successfully completing Airborne and Ranger Schools, he was assigned to 3d Battalion, 325th Infantry in America's Guard of Honor, the 82nd Airborne Division. Jim quickly learned his craft and excelled as a young leader.

When the call went out for volunteers for the 2nd Ranger Battalion, he quickly answered the bell and was selected to be in the initial cadre. As he describes it, "Being selected to join the Battalion was a wish come true. To be able to serve with outstanding officers, NCOs, and soldiers was a life altering adventure. Learning every day from all of them and being able to say I was one of them was something I will always cherish.

"As a young lieutenant, there were many good role models who each offered something unique and different. It wasn't hard to find good things to emulate. The most common attributes I most admired were professional competence, physical toughness, intelligence, willingness to lead others, and the need to live by a set of standards for all to follow. Many contributed to my education and helping to ensure I had a long and successful career.

"The two who had the most significant impact on me in my career were Buck Kernan and Lawson Magruder. Both took a keen interest in my assignments and took the time to offer advice and counsel when needed. Both impacted the way I developed over the years."

Jim spent three formative years in the Battalion. After serving for six months as Bravo Company's weapons platoon leader and company executive officer, he was promoted to captain and moved up to the Battalion staff. After distinguishing himself as a planner and operations specialist, he was selected for command of Charlie Company. This was quite an achievement as he had not yet attended the Infantry Officers' Career course.

The battalion commander, LTC Jerry Bethke, had tremendous confidence in Jim's ability to maintain exacting standards and continue the winning teamwork. Jim excelled, leading Charlie Company over the next year of intense training and emergency deployment exercises.

When Jim looked in the rearview mirror at his time in 2nd Rangers and how it positively impacted the rest of his military career, he offered the following: "There probably wasn't a day in the life of a battalion member that didn't provide opportunities for learning. Every day came with new challenges, and learning was a daily requirement. Some opportunities were easy and welcomed and others were hard but necessary.

"The process of preparing for each day's activities followed by immediate after-action reviews provided an intense learning environment, along with some painful moments. Everything we did was evaluated. We talked after every event or activity which contributed to learning. No one was exempt; officer, NCO, or individual Ranger.

"We lived and learned that there had to be a standard for everything.... we were not being evaluated against each other but against a specific standard which applied to everyone. This included physical training, military bearing and appearance, and tactical proficiency. All of us were required to meet the standard. Our exposure to the standards-based training philosophy was critical and formative. It hardened us to do the right things, even though it may not be the easy way. It cemented a need to have a clear standard in mind for everything we might do… and this was always on my mind when confronted with a difficult decision.

"I never forgot this is a firm leadership principle. Overall, my time with 2nd Ranger Battalion was the best education a young officer could receive."

Jim decided to stay in the Army after 2nd Rangers to help transform it over the next 25 years. He served in many elite units, including command of the 1-505th Parachute Infantry Battalion, 82nd Airborne Division, 3d Battalion 75th Ranger Regiment, and the 75th Ranger Regiment. He was also the G3 (Operations) officer for the 7th Infantry Division during Operation Just Cause in Panama. As a general officer, he had several key postings, to include Assistant Division Commander of the 2nd Infantry Division in Korea, and the Commander of the U.S. Army Military District of Washington. In each of these senior leadership positions, Jim focused on individual and organizational standards while genuinely caring for his soldiers and their families. This focus resulted in a foundation of trust that led to a climate of winning teamwork. Here is a rich list of the behaviors Jim has lived by as a military and civilian leader and family member:

- "Standards are for everyone. Without a standard, how do you know what is needed to complete the task correctly?
- The team is only as good as the weakest link. Everyone needs to be trained and proficient.
- Be an expert in your chosen field. Know the tools of your trade.
- Take care of your people. They will bring you success if you arm them with the right tools.
- Being hard doesn't mean yelling and treating your people poorly.
- Provide clear concise instructions.
- Stay focused on your mission. Remember, why "you entered the swamp."
- Never forget to thank those who work for you. It should be a daily opportunity. Don't let it pass by."

Jim Jackson's experience leading tactical units under the most demanding, dangerous, and complex conditions prepared him for the challenge of leading the military's emergency response to the attack on the Pentagon on September 11, 2001.

Major General Jim Jackson assumed command of the US Army's Military District of Washington (MDW) in July 2000. It is a large command responsible for supporting five Army installations in the National Capitol Region (NCR) and, at that time, Fort Hamilton in New York City. It had over 61,000 soldiers and civilians assigned to it and included the garrison staffs, the 3d Infantry Regiment (the Old Guard), the Army Band (Pershing's Own), and several other units. As the largest military component in the NCR, MDW had responsibility to coordinate joint and civil support in the event of a natural or manmade disaster or a terrorist attack on a military installation. It was a major responsibility that Jim and his staff took with seriousness and diligence. Prior to 9-11 they had conducted periodic tabletop exercises and coordination meetings with all the other stakeholders to include DC and county EMS and law enforcement agencies. Jim forged enduring trust and teamwork across diverse organizations during those exercises. His background as a team builder, planner, and operator resulted in detailed planning and preparation for what would be one of the most tragic days in our Nation's history.

When American Airlines Flight 77 flew into the Pentagon on that fateful day, every ounce of Jim's leadership skill was challenged under the most dangerous conditions over the next few days. His experience and "battlefield instincts," combined with detailed planning and preparation resulted in immediate response by his team in saving lives and reducing further disaster.

As Jim stated: "The support to the Pentagon has proven to be a model example of military civil authority cooperation. Our prior training and coordination paid off in spades. Our people knew everyone who would respond to a crisis within the NCR, and they were ready to execute. No major disagreements occurred concerning command-and-control. Each critical element understood how the system was expected to work and why unifying under a single incident commander was necessary."

The actions taken after the attack on the Pentagon on 9/11 and the lessons learned have been well documented. We will not detail them in this chapter. Rather, let us hear from Jim on his personal reflections and how his experience decades ago as a young company grade officer in the 2nd Rangers prepared him for the major leadership crucible in his career: "The idea of stepping up and doing what is needed without worrying about personal impact was very obvious."

"The concept of moving towards a problem area vice away from it happened that day. So many headed to those hurt or injured while disregarding possible personal injury."

"Individuals relied on their training to get out of some difficult, dangerous locations and others assisted those hurt inside the building so they could get out."

"Being a part of a larger team focused on the overall team objective was evident. As we learned early in the 2nd Rangers, everyone had to contribute to the team to get the job done. On 9/11/2001, no one was interested in getting recognition… everyone stayed focused on helping the team achieve its objective. Together many things were accomplished."

"We learned that success must be defined, and a standard established to provide a measurement to know if you are successful. We used this to help explain our plans and the desired outcomes to everyone we briefed or escorted at the crash site. When briefing senior leaders on a daily basis, it helped explain our purpose for specific tasks and actions and prevented unwanted distractions."

Jim commanded MDW for two more years and retired in 2003 after 32 years of distinguished service to our Nation. He and his spouse Nancy remained in Arlington, Virginia. Following retirement from the Army, he helped establish Battelle Memorial Institute's Office of Homeland Security in Crystal City. Battelle provides bio and chemical defense capabilities in defense of our Nation. Three years later, he became a Vice President followed by Senior Vice President for the Military Professional Resources, Inc (MPRI) International Group. At MPRI, he provided contractors, mostly former military veterans, in support of the war effort in Iraq and Afghanistan.

He was most proud of the caring support MPRI provided for their teams who were in harm's way a long way from home while contributing to the success of our military.

Jim continued to serve veterans from 2015-2020 when he was the Director for the Department of Defense Vietnam War Commemoration. The primary objective of the commemoration was to thank and honor Vietnam veterans and their families on behalf of our Nation for their service and sacrifice, with distinct recognition of former prisoners of war and families of those missing in action. The other objectives highlight the service of our Armed Forces and support organizations during the war; pay tribute to wartime contributions at home by American citizens; highlight technology, science, and medical advances made during the war; and recognize contributions by our Allies.

Jim was truly honored to lead the commemoration team. Once again, he built a team who trusted one another and understood and were passionate about their noble mission. "We helped thousands of veterans to come to some closure. Our team never lost focus on the individual veteran and their family. We shared many special moments with our veterans as we provided them a long overdue 'welcome home'."

Rangers Tim Grzelka, Jim McNeme, and Jim Jackson are exemplary examples of the positive impact that a transformative experience early in one's life can have on their future leadership journey. As plank holder contributors to the development of a foundation of trust leading to a climate of overarching teamwork in Bravo Company, they went on in their lives to pass on those winning ingredients to other organizations they led.

12. RELATIONSHIPS

"It takes commitment and time to typically forge friendships. But if you have been cold, wet, tired, and hungry together, those experiences and memories accelerate camaraderie."
— CPT Lawson Magruder

From left: John Welgos, Mark Lisi

"RAVEN!" WELGOS BARKED ON THE RADIO.

"What's up, John?" Raven replied ('Raven' was co-author, Fred Kleibacker's radio call-sign).

"We have a situation at the front gate, can you come up?" John requested.

"Roger, be there in five," Raven replied.

It was the summer of 1993. Raven jumped into the rental vehicle and sped uphill from the training site to the main entrance of the former school for handicapped kids. The school had been closed and abandoned for years. It was a perfect training venue for the secretive, counterterrorism (CT) unit.

The sprawling compound of two-story brick structures was secluded within a deep forest, far from roads. The school was directly across Interstate 295 from the small village of Bordertown City, New Jersey. Bordertown is famous as a Revolutionary War battle site just south of Trenton, New Jersey, and about an hour south of Edison, New Jersey, which is just south of New York City.

It was about 2 A.M. and the secretive, US military group had been conducting live fire, CT exercises most of the night. They were running building assault evolutions to clear and secure buildings while dispatching, role-playing terrorists. The CT unit was being supported by the Edison, NJ, Police Department's (PD) Special Weapons and Tactics (SWAT) Unit led by Lt. John Welgos.

Raven pulled up and got out of the vehicle. There was a group of Edison SWAT members facing a young woman holding a screaming baby in her arms and about a half dozen other citizens of Bordertown City complaining about the noise.

"Ma'am, I'm so sorry the noise scared your baby. We're winding up." Raven heard John say as he approached.

"We don't understand what's going on. It sounds like a war. Are we being invaded?" The young woman sobbed. The small crowd echoed her complaint.

It had been a breezy, summer night. It was cool and the humidity was low. The result was that the loud noises were carried across an empty I-295 by the wind towards the small village of Bordentown, NJ. To the 2,500 people just a quarter mile across the highway, it sounded like a war zone. The cacophony of war noises was terrifying. The reports from rifle and pistol fire, blasts from explosive door breaching and exploding concussion grenades, known as "flash bangs," had started just after 10 P.M. and had been going almost non-stop for the past four hours.

"No, ma'am, we are conducting SWAT training." John replied.

"I don't understand. What's SWAT training? Why are you doing whatever you are doing here? Are you planning on attacking us? Are we in trouble? How much longer is this going to last?" The rapid-fire questions left the young women breathless, her voice cracking with fear and frustration as tears dripped from her eyes. She pulled her crying infant closer to her body, coddling him in her arms against her chest.

"Ma'am. I can assure you that no one in Bordertown will be hurt. I'm Lt. John Welgos from the Edison PD. I'm also the SWAT Commander, these men behind me are my team. We are conducting training that I personally coordinated with your police chief a few months ago. We were approved by the city council to use this venue and conduct training. I'm sorry they did not warn you," John said gently.

She didn't reply, just looked at him with a confused look on her face. John's soft voice and sincere compassion for her, her baby, and the town seemed to be breaking through. John's voice got a little softer, more focused. He spoke slowly from the heart as he addressed the small crowd.

"Folks look, this training is necessary to protect the public. We dedicate our lives to the service and protection of our citizens. We live in a very dangerous world. As you may recall, terrorists attempted to blow up the World Trade Center earlier this year. While we're not in a declared war, make no mistake, we are in a war against a very dangerous enemy who wants to destroy America and our way of life. The men down there training are the tip of the spear of that war.

"They need to use realistic locations. They need to train daily. Their skill sets require years of training to hone and maintain so they do not make mistakes when the day comes that they must use them. Those men are the bravest men in America. They are the real patriots. They are of the same ilk as the colonists who fought in the battles of Trenton and Bordertown and who braved the ice and cold to cross the Delaware to turn the tide of a failing war and to save our nation on Christmas Day in 1776.

"We are sincerely sorry we disrupted your evening and night, but it can't be helped. I hope you all will understand."

Amazingly, her baby had stopped crying, even though the explosions, rifle reports, and flashbangs had not stopped during the short exchange. Without warning, as tears streamed down her cheeks, the woman closed the distance with John and gave him a strong, emotional hug with her baby in between them.

"I'm so sorry, I didn't know. God Bless you all, and all that you do. I guess we can survive one night with a little noise. It's the least we can do," she said softly.

John joined the Army in the spring of 1971, five days after his 17th birthday. He came from a family steeped in paratroopers. His dad was a paratrooper with the 82nd Airborne Division in World War II, making all four combat jumps. His uncle John was also in the 82nd as a combat medic. His uncle Phil was an officer in WWII. John knew at a very young age he wanted to go into the military and make a career out of it. He followed in his father's and uncles' footsteps. After completing Basic, Advanced Infantry Training, and Airborne School, John arrived at Ft. Bragg, NC, and joined the 82nd Airborne Division. In fact, he went to the exact unit his father had served with during WWII as a multi-generational paratrooper.

John quickly demonstrated he was someone special. As a young Specialist (E4), John's motivation and performance won him "Soldier of the Month" and then "Soldier of the Quarter" awards for the Division during his first year. The reward was spending a day with the Commanding General of the 82nd Airborne Division. John did his homework on the General and learned he had attended the United States Military Academy at West Point and that he had a storied career as an Airborne Officer. John arrived and was briefed by the General's Aide de Camp on the "do's and don'ts" while spending the day with the General.

They got into the General's car and the General abruptly asked John, "What are your plans, young man?" John was ready. John recalls answering, "More than anything, Sir, I want to go to Ranger School." What he really wanted was to go to the Special Forces, but he couldn't because he was too young.

The General said to him, 'Son, you know we have a mighty fine Recondo school here at Fort Bragg."

John replied, "Yes Sir, I know, but the way I see it, why go to a Technical College if you can go to West Point?" With that, the General grinned and turned to his aide, and said, "See that the good specialist gets orders to Ranger School." So, John got orders, at 18 years old, for Ranger School.

John graduated as the Distinguished Honor Graduate out of his Ranger class and returned to his unit as the only "Tabbed," E4 Ranger. A few months later, he was promoted to Sergeant E5, with barely 20 months in the Army.

John's potential was soaring. He had built excellent relationships with his chain of command. He demonstrated competence and leadership. So much so, that they encouraged him to go to West Point. John went to the West Point summer preparatory program at Ft. Belvoir, VA, and right before he was to begin his Plebe (freshmen) year, a routine physical discovered he had a high frequency hearing loss from his time in the 82nd. It disqualified him from attending. He still had time on his contract, so the Army sent John to Recruiting Command.

It was a mandatory three-year tour, so John re-enlisted and was sent close to his hometown in New Jersey as a recruiter. Only 20 at the time, he excelled at his job because John was a natural relationship builder, as well as being close in age to the future recruits. He was also a fierce competitor and exceeded the recruiting goals set by his station commander. Shortly after his first year there, an article in the Army Times caught John's eye about the Army standing up the 2nd Ranger Battalion at Ft. Lewis, WA. They were looking for NCO volunteers. Still too young to go to Special Forces, John called the phone number listed in the article.

He recalls, "It was Friday, so I made the phone call. My station commander said I was wasting my time; nobody gets out of recruiting. After a short wait, the secretary transferred the call to Command Sergeant Major Morgan, the 2nd Ranger Battalion CSM. He asked me a couple of questions. 'What's your name and rank? Have you been to Ranger School? What other training and qualifications do you have? Can you be in my office Monday morning in your class A uniform with your 201 Personnel File?'

"I told him yes and immediately went down to pick up my personnel record. I booked a round trip flight from Newark, NJ, to Seattle, WA, for the following day. When I got there, I rented a car and drove down to Ft. Lewis, about an hour south of Seattle. I figured out where the Ranger Battalion was located on post and identified the Battalion Headquarters.

I got a hotel room near the base, shined my boots, polished my brass, and set up my Greens."

"I showed up in front of the Sergeant Major on Monday morning at the prescribed time with a perfect dress uniform. It started out routine. Then suddenly, he looked up at me and he asked, 'Did I talk to you on Friday morning?' I said, 'Yes, Sergeant Major.' He asked, 'Where were you?' I said, 'New Jersey.' He said, 'How'd you get here? On a military flight?' I said, 'No, Sergeant Major, I bought a round trip ticket to come out for the interview. My command said that there was no way that I would be released from my recruiting assignment.' The Sergeant Major dismissed my comment with his hand and replied, 'Don't worry about what they say. Go back home and start packing, because you'll be getting orders for the 2nd Ranger Battalion.'"

John was assigned to 2nd Platoon, Bravo Company as a Sergeant (E5) Squad Leader. In short order, he was promoted to Staff Sergeant. John described his time in the Battalion as, "My Ranger Battalion time was simple. We didn't have any secret tactics or weapons. What we did there was we literally mastered the basics of soldiering and being light infantry. That's what we did. We mastered it.

"There's an old saying, 'Amateurs practice until they get it right. Professionals practice until they don't get it wrong.' And literally, that's what we did.

"And I was so proud to be there. I was so proud to be a squad leader. I had a great squad. I had great team members."

In the spring of 1977, John came down on orders for Germany with no unit assignment. When he arrived at the Replacement Depot, the 1st Battalion, 10th Special Forces Group had a representative there asking any Ranger that came out of one of the two Ranger Battalions if they wanted to become a Green Beret. John finally was going to be a Green Beret. He went through the training and stayed there until 1979. John made the tough decision to leave active duty because he had a confirmed slot at the

NJ Police Academy and an offer with the Edison Police Department upon graduation.

Later, John joined the 11th Special Forces Group in the Reserves, eventually becoming a part of the Individual Ready Reserves (IRR) as a Special Forces senior NCO.

After John finished the police academy, he became a patrol officer, eventually working his way up the ladder to become the Narcotics Unit Commander and the SWAT Team Commander. This brings us back to the beginning of John's story.

In 1988, John had run into another former 2nd Ranger Battalion Ranger who was now part of the Army's secretive Counterterrorism (CT) unit.

In John's words, "I was at Fort Dix. I ran into another former 2nd Ranger Battalion guy on the tarmac getting ready for a jump; he was from C Company and was then serving in the Unit. We started talking and catching up. He asked me if I could help them set up some training. I said, sure. So, I wound up creating training areas in New Jersey, New York, Philadelphia, and Atlantic City for his organization.

"I became a liaison in the Northeast. They would tell me where they wanted to train, and I would set it up in all these big places like New York City. I brought other law enforcement organizations together as well. It all started with two runaway Rangers and ended up with me hosting these guys multiple times between 1988 and 2001. Of course, that all ended after 9-11.

"We were finding training venues for them, they would come and look at them, meet and coordinate with the owners of the venues, and set up schedules typically about six months out. They did live fire, live explosive breaching, helicopters, sniper shots from helicopters, sniper shots between buildings, rappelling from 50 story buildings, all the cool stuff. Our job was to assist in the coordination, planning, building relationships between organizations, and then maintaining a perimeter primarily for safety and to keep civilians out. The relationship became very close. We were invited annually to visit their compound for sniper and SWAT competitions."

For more than 10 years, John was the east coast police liaison to that organization. Through his relationships with multiple law enforcement departments in New Jersey, New York City, Atlantic City, and Philadelphia, John was the go-to guy.

John was able to retire from the Edison PD after 9-11 and then deployed to war zones all over the world as both a protective security detail for U.S. government officials and as an active-duty, Special Forces NCO. He retired after his eighth trip to Afghanistan where he had been seriously wounded and suffered a traumatic brain injury (TBI). It was time to hang up his spurs, so to speak.

John spent 32 years total time in the Army, both on active duty and in the reserves, deploying eight different times with Special Forces to Afghanistan and four times as a contractor to Iraq and Gaza.

Today, John is a President of a National Veterans Motorcycle Club serving vets up and down the East Coast. He also spends his days hiring himself and his tractor with a backhoe out, mostly to help elderly folks with inexpensive landscaping and excavation projects.

John is all about personal relationships. He has built relationships all his life and actively keeps them. Perhaps it is his personality, but it is also because he is competent, and he can be trusted. It is this foundation of trust that builds relationships. As the old adage suggests, 'Say what you mean, mean what you say.'

Of course, John sees it a bit differently, "My most valuable possessions are my word and my intent. I've been diagnosed as having a rescue complex. I want to contribute in whatever capacity I can, as long as I can; because I've been blessed in so many ways in my life."

* * *

This brings us to another incredible Ranger who builds lasting relationships—Mark T. Lisi. John and Mark are "brothers from a different mother." Completely different in personalities and styles, but their ability to bring people together and organize an event for a successful outcome, is very similar.

It was 2004, Fred Kleibacker's administrative assistant came to his office and peaked her head in and said, "Hey, there's a Colonel Lisi on the phone asking for you?"

Fred smiled crookedly with a tilted head and said, "*Colonel* Lisi?"

She said, "That's what he said."

Fred said, "Send him through. Thanks."

Fred hadn't seen Mark Lisi since 1978 when they were young Sergeants in Bravo Company 30 long years ago. Fred thought to himself, 'Lisi is a *Colonel?* How'd that happen?'

Fred's phone rang, he picked it up and said slowly, "*Colonel?* Lisi?"

"Freddy! Mark T at the jump TOC," he bellowed.

Fred said, "The last time I saw you, I don't think you had any eyebrows!"

It was the last word Fred got in.

"Freddy, I was in Afghanistan last week and I had this thing. I fell down. My chest hurt," he blurted.

Okay, it wasn't Fred's last word.

"What thing, Mark?" Fred asked.

"Oh, it's nothing. They say I might have had a heart attack. But you know, you can't keep a Ranger down," he said.

"Mark, what do you mean by 'I might have had a heart attack?'" Fred replied.

"Freddy, you know, your heart stops! Are you stupid?" He angrily retorted.

"It's not the first time someone has called me that. Where are you, brother?" Fred asked.

"Walter Reed Army Hospital," he said.

"How did you find me?" Fred asked.

"Freddy, I got resources, I'm a Colonel for crying out loud," he announced.

"I heard that…what's your room number?" Fred inquired.

"Room number! I don't have a room! They got me in a damn closet!" he roared.

"Okay, how long are you going to be in your closet?" Fred chuckled.

"What, do you think I'm a mind reader too?" he snarked.

"You're boring me now, Mark." Fred quipped.

"I don't know, they got to do tests and stuff, Freddy." he puffed.

"I'll be up tomorrow, what ward are you on? Make sure you put me on the visitor list," Fred ordered.

"What, you don't think you're not already on it? That's why I called. I'd really love to see you," he declared defiantly.

"Me too. Get some rest, brother. Focus on getting better," Fred advised.

"Okay. I got to go. I think the nurse likes me," he joked.

Click.

This might all be an old, highly inaccurate memory, but it isn't far from the truth. Ask anybody who knows Mark T. Lisi.

Mark is the son of a legendary Airborne Ranger and Special Forces Command Sergeant Major. After graduating high school, he wanted nothing more than to attend his cherished Virginia Military Institute (VMI). He got his wish, and in August of 1972, found himself in Hell Week as a Freshman "RAT." For Mark, the challenge wasn't enduring the week-long hazing or the military discipline, it was academics. At the end of his freshmen year, he realized he wasn't ready for college.

In his own words, Mark relates how he decided to join the Rangers, "I was back home at Ft. Lewis, WA, washing dishes in old Bravo Company Rangers' mess hall contemplating my future. Later that summer, I broke my leg on a sky diving parachute jump. My work world got very small and very dark, and I decided it was time to join the Army. Frankly, because my dad was a Ranger, and this is going to sound very trite, all I ever wanted to be, was an Airborne Ranger."

Mark joined the Army in November of 1974 on a three-year unit of choice contract, with the 2nd Ranger Battalion. Following the usual pipeline, Basic, AIT, and Airborne School, Mark showed up at Bravo Company in early April 1975 as a PFC (E3). He was warmly welcomed to the company by the First Sergeant, 1SG Bill Block and the Company Commander, CPT Lawson Magruder. He was assigned to the 3rd Platoon. Like the rest of us, Mark had a lot of growing up to do. We were all untested, untrained, and largely incompetent privates.

Fred Kleibacker: "Most of us were intimidated, didn't know what was expected of us, and uncertain if we could cut the mustard. The officers and NCOs were the most professional soldiers we had ever been exposed to in our short time in the Army. For the next three years, we were melded and forged into a deadly fighting force. In Mark's case, he matured from an obnoxious, loud-mouthed private into a highly competent, slightly less obnoxious, young Sergeant."

Mark fondly remembers, "I did my three years like all of us, and like all of us, experienced considerable personal growth in those three years. I learned that I could endure, I could suffer and survive. I learned I was way more capable than my 1.6 GPA from VMI would indicate."

At the end of his three-year tour, he decided to go back to college. He applied to Eastern Washington University (EWU) in Spokane, WA, with glowing recommendations from several officers, including CPT Magruder, who was now the Battalion S1, and CPT Jim Dubik, who was the current Bravo Company Commander.

He was accepted immediately. EWU had an ROTC Department and were delighted to have the young Ranger as part of their program. Mark received his BA in English with a focus on Poetics and Writing. He graduated as the detachment's Distinguished Military Graduate, representing the top 20% of the best ROTC Cadets across the nation. He was commissioned as a Second Lieutenant in the Infantry, in 1984. After graduation he attended the Basic Officers Leadership Course at Ft. Benning, GA, and received his first active-duty assignment to Alaska as an Airborne Infantry Rifle Platoon Leader.

After Mark finished his tour in Alaska, he returned to Ft. Lewis, WA, and would serve two tours as a Company Commander in the 9th Infantry Division at Ft. Lewis, WA. In between commands, he served in various Staff Officer positions at the battalion level. He left Ft. Lewis for a tour in Korea at the Joint Security Area, patrolling the Demilitarized Zone (DMZ) between North and South Korea.

Mark returned and deployed to the First Gulf War. Now with 15 years of active-duty service, Mark decided to take advantage of a Voluntary Separation Program that the Army was offering following the Gulf War. He chose to continue to serve as part of the Individual Mobilization Reserves on a part time basis. This gave him the opportunity to go back to school and acquire a teaching certificate.

For the next few years, he taught middle school English while continuing to serve in the military as a reservist. After 9-11, Mark was eager to get back into the fight. He was able to secure an Individual Ready Reserve (IRR) position once he was promoted to full Colonel. He would finally get deployed to Afghanistan as an advisor at the Office of Security Cooperation helping to build the Afghan Army.

Mark finished his one-year tour and while on his way home, had the heart attack that began this story. Mark went on to serve a few more years and officially retired from the Army in 2006, after 32 years of service to his nation. Mark still needed some adventure, so he did a short tour as a contractor in Iraq running a security contract protecting Iraqi judges and lawyers. But time was taking its physical toll, it was time to seriously call it quits. He came home and eventually moved to Reno, NV, to fully retire.

Mark spoke of the importance of his time in Bravo Company, "Virtually every guy in that company, if you ask him, in his own way, would tell you that the time they spent in B Company was to define or shape who they became as adults."

It is a profound statement. Most of us didn't know how to understand it until Mark led the effort to get us all back together. But it wouldn't be for another 13 years.

In 2019, Mark began an effort to assemble the original members of

Bravo Company. It was a herculean effort to track down so many Rangers who served between 1975 and 1978 in the original Bravo Company Rangers. Mark was already in touch with a lot of us, but not everyone. He assembled lists of names, numbers, emails, and addresses. He had done his staff work. He made sure LTG (Ret) Magruder, co-author of this book and our first Company Commander, was in the loop and supported his effort to reassemble as many Bravo Company Rangers as possible.

Fred Kleibacker: "That's when I got a call from Mark out of the blue. We hadn't talked since I visited him at Walter Reed, 20 years prior.

"Freddy, I'm organizing a reunion. General Magruder has approved it. I can put it together here in Reno. Everything is turnkey. We can stay at a casino, it's got everything we need, plenty of restaurants, entertainment, and I can get them to comp a hospitality room for us to gather. I'll arrange a banquet room for a semi-formal dinner."

He and Fred started to catch up. He told Fred that he was on the Board of the Ranger Scholarship Fund. The Fund gave merit-based scholarships to children and spouses of former Rangers of the 75th Ranger Regiment. After talking, Fred said, "Let's raise some money for the fund during the reunion."

"Okay, you're in charge of fundraising!" he ordered Fred.

Great! I had to come up with an idea. Thank goodness my wife is a better entrepreneur than I am. We were vacationing in the Cascades when we ran across a local whiskey distillery. They had blended casks of bourbon that we could label without incurring huge customizing costs. At a cost to us of $35 a bottle, we could sell it for $70 with a couple of cool, etched glasses with the 2nd Ranger Battalion scroll, as a set. All the net proceeds would go to the fund. We made almost $5k after expenses for the scholarship Fund. But that is a side story. The real story is the miracle of the first ever Plank Holders of Bravo Company reunion, orchestrated by COL (Ret) Mark T. Lisi. We held a second one in 2023 and again raised around $4,000.

Because of Mark, we now have a Bravo Company Plank Holder legacy—The Major General Eldon A. Bargewell Scholarship Award. Named after our famed 3rd Platoon Leader, Eldon A. Bargewell, who sadly passed away in 2020. It is a permanent scholarship award within the Ranger Scholarship Fund from the B-2-75 Plank Holders.

So, what was this itch we all wanted to scratch? Mark wrote, "The leadership and the training routine served to mold soldiers at the elite level. They could do anything, correctly, the first time. Learning to endure hardship and suffering while still accomplishing the mission was the core task. Make no mistake about this; it is a learned skill. That magic nexus where skill, desire, and endurance all meet.

"Bonding played a big role in the early days. Much of the bonding is deeply rooted in the suffering. Cold, wet weather, and road marching will test everyone. These sorts of bonds tend to last forever. These are the men who validated the original Charter. Good men, solid men, young men who could be relied upon to do the hard thing, the right thing. Wrap these men in the Ranger Creed and you have a complete package."

Mark continued, "In short, all the original Rangers in Bravo Company will tell you in their own words that their time in the Battalion would go on to define who they became as adults and professionals. It was a magical time for a young Ranger and a gateway to a successful life. Many might say I wear my "Rangerness" like a coat of armor, perhaps I do, well, I do."

When the Officers, NCOs and enlisted men who served in Bravo Company during those early years were reacquainted during the first reunion it was as if they had never been apart. They were all intensely proud to be part of the first, post-Vietnam, 2nd Ranger Battalion. Airborne Rangers who set the standard that would remain within those units for the next 50 years.

From the first stanza of the Army NCO Creed: "No one is more professional than I. I am a noncommissioned officer, a leader of Soldiers. As a noncommissioned officer, I realize that I am a member of a time-honored Corps, which is known as "The Backbone of the Army." I am proud of the Corps of noncommissioned officers and will, at all times, conduct myself so as to bring credit upon the Corps, the military service and my country, regardless of the situation in which I find myself. I will not use my grade or position to attain pleasure, profit, or personal safety."

MSG (Ret) John Welgos and COL (Ret) Mark T. Lisi, two extraordinary men who knew how to build lifetime relationships to make a difference in the future.

13. SENSE OF PURPOSE & DIRECTION IN LIFE

Time and again, service to something greater than self has proven to provide the "why" and azimuth for the rest of your life.
—CPT Lawson Magruder

From upper left: Joe Picanco, Ed Land, Joe Wishcamper

1972 MESA, ARIZONA. JOE PICANCO HAD BEEN RUNNING TOO LONG: RUNning away from a terrible childhood after his father died when he was 5; running away at the age of 16 from an abusive, alcoholic family unit; marrying at 18 and fathering a child soon thereafter; and going from job to job until he became a car painter to support his young family.

Now he wanted to stop "wandering in the desert" and find direction and greater fulfillment and purpose in his life. So, he went down to the local Army recruiter to talk about options. Even though he was afraid of heights, adventure, and the life of a paratrooper—who happened to make $55 more a month than other enlisted soldiers—appealed to him. One recruiter said he was too scrawny and had bad eyesight and should look at another specialty.

But an airborne Ranger recruiter in the same office said he could get him into the Rangers through the enlisted Ranger option. Joe immediately left a job that paid $3,000 a month for one that paid a "whopping $400 a month" and enlisted. Five months later, he was headed to basic training at Fort Benning, Georgia. That decision would change Joe's life forever, just like it did for the other veterans we interviewed. For the next 30 years, Joe would have a clear purpose as a proud member of a family he never had before: the United States Army.

Joseph "Joe" Manuel Picanco was born 1952 in Hanford, California of Portuguese heritage from parents who migrated there prior to his birth. After a very rough, abusive childhood too painful for him to describe, he decided to break the cycle and change the azimuth in his life and make something of himself.

After he completed basic and advanced infantry training and airborne school, he found himself in Ranger School Class 2-74 with a cohort full of college Army ROTC cadets. Although he lacked the leadership training of other students, he listened intently and learned from the cadre. He successfully completed one of the most challenging leadership courses as a young, enlisted soldier. His Ranger tab was proudly pinned on his left shoulder in October 1973, and he was promoted to Specialist from Private E2. Joe was extremely proud of his accomplishment, but it came at a cost as his first wife refused to follow him to Fort Lewis for his first assignment in the old B Company 75th Rangers. They divorced shortly thereafter.

Joe found not only excitement and adventure in the Rangers, but also the family unit that he had been looking for since his childhood. Shortly after his arrival, he was made a squad leader in the second platoon. He immediately bonded with his fellow squad leaders particularly Don Hibbs who a year later would be the best man in his wedding.

In Joe's words: "I think the reason why the Army was such a great thing for me and particularly the Rangers, it's that family piece. I did not have that growing up. The Army became my family. And we took care of each other. We worked with each other and helped each other through difficult times. That whole family unit that I was lacking came together, particularly in Bravo Company. We bonded because of the difficulty of tasks that we accepted in terms of our training. And we succeeded at them, and we pulled together. We did not tolerate failure in anyone. We worked hard to help our young soldiers succeed in every endeavor."

An example of Joe and his squad being given an opportunity to succeed and ultimately excel, happened early after the formation of the Bravo Company. It occurred during a challenging squad level event called the "march and shoot." It was an event consisting of a timed 15-kilometer road march, culminating in scored shooting on a rifle range.

Joe recounts: "My squad was well prepared for the march and shoot. That morning, the planning, the movement, all of that went extremely well but I hate to use this word, we failed on the range. We missed perfection by one point. In other words, we were one target short of being right on the money. One of my team leaders Sergeant Shannon looked at me and I could see it on his face, that disappointment. Sergeant Erickson also looked at me and I could see it on his face, that disappointment.

"CPT Magruder came to me and said, "Sergeant Picanco, what do you want to do? Do you want to do it again right now or just take the score?" We immediately took the opportunity to walk the 15 kilometers again, and to achieve a perfect score that same day. That opportunity to demonstrate that the squad had the capability of maxing that score was incredibly motivating. You cannot believe what it did for my team and squad. It taught me a lesson that not every failure has to end that way. You can turn it into success with something additional… either guidance, training, or motivation."

Joe had many memorable developmental moments during his two and a half years in the battalion: "As we worked through our accelerated program from individual, to team, to squad and platoon, and eventually up to company level training, it was amazing to watch our soldiers looking to their leaders for training in how to do things to standard. And then to admire them doing those things to standard. And to watch my fellow squad leaders and other leaders focus on mission success and personal success in terms of understanding and accomplishing their mission. That was an incredible experience.

"The leaders in the battalion, from LTC Bo Baker on down, truly led from out front. If we were cold, wet, tired, and hungry, so were they. There was respect for one another regardless of rank. We understood rank, but we also had respect for one another regardless of our rank. We included one another in the planning and decision-making process and fostered ideas from one another.

"I decided to join the 2nd Ranger Battalion because of what I heard. First, it was supposed to be the toughest fighting unit in the United States Army. No other infantry battalion would look like this. Second, it would be disciplined and have the highest moral conduct. We would not be hoodlums, or we would be disbanded. Well, when we went to cadre training at Fort Benning, Georgia, four of us did get into a fight—me, SSG Demoisey, SGT Shook, and SGT Rekasis.

"We got into a fight, but quite frankly, we were protecting ourselves. We did not start the fight; we were attacked at a club in Columbus. The next day I stood in front of LTC Baker and CSM Morgan, and LTC Baker told me that if were back at Fort Lewis, I would be thrown out of the battalion. I was white as a sheet of paper. I could not believe they would throw me out of the battalion. Thankfully, I was not.

"One of the proudest moments I have ever had was 18 months later at COL Baker's farewell. He came up to me and he said, 'You remember that time at Fort Benning?' I said, 'Yes, sir, I do.' And he said, 'I'm glad I didn't do that!'"

That is a story about second chances; about belief in the potential of another; about resilience; and about redemption.

Joe loved his time at Fort Lewis, not only because of his service in the 2nd Ranger Battalion but it is where he met and married the love of his life, Kathi. She had just completed her enlistment in the Women's Army Corps (WAC) and was ready to continue her adventure with Joe for the next 27 years in the Army. Kathi's story can be found in Chapter 15.

Joe's assignments as a Staff Sergeant and Sergeant First Class would take him to infantry units in Hawaii, Fort Benning, Korea, and Italy. At each posting, he inevitably ended up as an acting platoon leader because the company was short lieutenants. Because of his high standards, he became a "fireman of sorts" in dysfunctional units. He became the standard bearer and role model for others more senior to him. His example effected positive change in every platoon he led.

Joe returned to the 2nd Ranger Battalion as a Master Sergeant and soon took over as First Sergeant of Alpha Company after serving as Operations Sergeant. He was back "home" and loved being around Rangers and donning his black beret again. But it was to be short-lived because he soon came out on the list for the Sergeants Major Academy.

He reluctantly left the battalion and left Kathi and their two children in Olympia, WA, thinking he would return upon graduation. After the Academy, he had a short stint training ROTC cadets at Fresno State University until he was promoted to sergeant major (E9). Joe spent the next 15 years as a Command Sergeant Major in some particularly challenging combat ready tactical and training units: 1-509th "Geronimo", the opposing force battalion at the Joint Readiness Training Center; 3-325 Airborne Combat Team, SETAF Lion Brigade in Vicenza, Italy; 1-508 Parachute Infantry and 7th Army Training Command in Germany. He culminated his career as the Command Sergeant Major of the U.S. Army Cadet Command, responsible for all junior and senior ROTC detachments in the Army.

When Joe looks back on his 30 years of service, he states: "After my formative years in 2nd Rangers, I was always focused, in every unit in which I served, on training to standard, raising standards, and insisting on those being met. I and other 2nd Ranger plank holders truly planted seeds for the rest of the Army.

"I also tried to maintain a positive attitude, no matter how challenging the conditions. I learned it at a very young age in a dysfunctional family. I was always trying to make something positive from a negative. Because if I didn't, I would just wall it out or run from it, or just get into a mental state where I'm depressed all the time. So, at every opportunity, I would either look for it or create a positive atmosphere wherever I could, wherever I was."

Those who have served with and know him would resoundingly state that Joe Picanco represents the positive, servant leader.

When Joe retired from the Army in 2003, he and Kathi returned to the beautiful Pacific Northwest and have resided in Olympia, Washington near Joint Base Lewis McChord (JBLM) for the past 21 years. Their son Jason followed in his parents' footsteps and served honorably as a military policeman for 20 years, to include service in Iraq and Afghanistan where he was wounded and earned the Purple Heart. He continues to serve as a DOD security force supervisor at JBLM.

In "retirement," Joe has continued to fuel his purpose in life to serve others. He has been a leader in his church, providing security at church services. He has also been highly active in the Independent Order of the Odd Fellows where he served as Grandmaster. Now he serves as the state representative to the Sovereign Grand Lodge at the International level. Its noble tenants go back to its founding in Great Britain in 1784: to visit the sick, to relieve the distress, to bury the dead, and to educate the orphan. Joe has been a key leader in the organization, modeling for others the values of integrity and duty.

As the overseer of many business accounts, he has ensured that when fraud and abuse have been encountered, the offenders have been held accountable. Even as a volunteer in nonprofit outfits, Joe continues to ensure just as he did throughout his career that standards are established and enforced!

* * *

Unlike Joe Picanco who decided to remain in the active Army for a career, Ed Land and Joe Wishcamper decided to leave the Army to pursue other professional endeavors. Both regretted not staying on active duty, although Ed did have a fulfilling career in the Army National Guard. Here is his story:

John Edward "Ed" Land Jr was born and raised in Winchester, Tennessee. His dad and an uncle served in the Army during the Korean Conflict. He remembers when he was five and first saw soldiers on a rifle range. He was inspired by their uniforms and thought that maybe someday he could be like them. Throughout his childhood, he was in scouting. He loved everything about the uniform, the adventure, the outdoors, and the knowledge he gained as a Boy Scout. He became an Eagle Scout in 1970. It was not surprising that when he enlisted in 1973, it would be for the four-year Ranger option.

As an 18-year-old, Ed successfully completed Airborne and Ranger training and was assigned to a training battalion at Fort Benning, Georgia only few hours away from home. Because he was not adequately challenged, he soon volunteered for the newly activated 2nd Ranger Battalion. He arrived at Fort Lewis in early 1975, shortly after the battalion was activated. Trained as an infantry mortar specialist, he was assigned to the Weapons Platoon which had 60mm mortars and 90mm antitank recoilless rifles.

For the next year, he and his team became experts on antitank gunnery and supported the rifle platoons during grueling training, to include the battalion's extraordinarily successful ten-day certification exercise. By the end of the year, Ed had been promoted to sergeant and was the antitank section leader. He was now an experienced, highly respected junior NCO in the company with more advancement on the horizon, but it was not meant to be... because as Ed describes it; immaturity and impulsiveness got the better of him.

Ed was upset with the "big Army" because it had promised him along the way that he could remain on active duty and be assigned as a paratrooper to the 101st Airborne Division at Fort Campbell, Kentucky and be close to home. Unfortunately for Ed, the 101st had recently converted from an airborne to an air assault division, so there were no positions for paratroopers. So, Ed decided, against the advice of others, after 18 fulfilling months in Bravo Company, he was getting out of the Army after his enlistment ended in a year.

He requested reassignment to the 3-39th Infantry Battalion in the 9th Infantry Division which was a few doors down from the 2nd Rangers. For the next year, he took all he had learned in the Rangers and positively impacted the soldiers he led in his infantry squad. He focused on tasks, conditions, and standards and made field training realistic and exciting for his soldiers.

He built esprit and competence and confidence in them. His squad became the model of trust and teamwork for his company. All the way to the end of his enlistment, Ed was the model "Be, Know, Do" NCO.

After Ed left active duty, he returned home to Winchester, expecting to soon find the same fulfillment he experienced in Bravo Rangers, but it took a while. He worked in his mom's new restaurant for 18 months, but his mom soon found out that being a good cook did not make for a successful restaurant. Ed soon gravitated back to his comfort zone and joined the Tennessee Army National Guard.

He was quickly recognized for his leadership skill and was sent to Officer Candidate School (OCS) and was commissioned an officer in the field artillery branch. For the next 24 years, he served in a variety of key positions on state active duty, rising to the rank of LTC. In each assignment, whether it was in an artillery unit or at state headquarters, he brought his focus on standards, training management, and realistic training to his part time soldiers. He was admired for his competence and patient, positive leadership during a transformative time for the active, guard, and reserve Army. While in the Guard, he used his GI Bill benefits and earned a degree from the University of Alabama at Huntsville.

Here is what Ed took from his time in the 2nd Rangers that served him well as a leader in the National Guard: "One of the big takeaways from my time in the battalion was that we were constantly busy training. From the time we got up in the morning until we got cut loose in the evening, our day was planned. We had very little idle time. Everything we did was against a standard.

"The relationships we formed many years ago are so important. Being reunited with my platoon leader, Marshall Reed, and platoon sergeant, Roy Prine, after many decades has been quite special. They were positive role models during my formative years.

"I learned that if I wore the uniform with a Ranger tab on it, I needed to "Lead the Way" at every endeavor. In the Guard, I always felt I had a bullseye on my chest. That more was expected of me. I was up to the challenge.

"I learned how to be persistent, patient, and respectful as a leader. You can inspire others to change their behavior by demonstrating proper behavior for them."

After Ed retired from the Tennessee Guard, he took his expertise in information systems and became the domain manager for the Franklin County Board of Education. For the next two years, he inspired the younger IT specialists to perform their tasks to standard and modeled integrity, duty, and hard work for them.

Ed is now in retirement, but it has not been easy. After being heavily involved in scouting since he was seven years old, his world was rocked a couple of years ago when he was falsely accused of inappropriate contact with a young man.

With Ed's approval, here is his account of what happened and how he is coping with this setback: "Four years ago, during the sad period of all the accusations of child molesters in the Boy Scouts, one of my scouts turned me in without me ever seeing the charge. I just get this phone call from our council executive who says, 'You have been named in this big lawsuit, and we are having to put you out. You cannot wear the uniform. You cannot attend Scout meetings. You cannot make district meetings.' And that cut me to the core because I had been a scout since I was a kid.

"In my case, there was a scout who came from a challenging family who had been unruly and hard to manage for some period of time who finally showed up one night and actively participated and was doing good. I walked up to him, and I put my hand on his shoulder, and it startled him. He said, 'Don't touch me!' and I apologized and walked away. It was not inappropriate touching. The boy stayed in scouting for three more years and then I was included in this huge lawsuit that does not even list the victims. I hate this misuse of lawsuits for financial gain when it takes away from those who have genuinely suffered abuse.

"The event occurred in 2003, and I was released from scouting in 2021."

Despite his issue with BSA National, Ed remains committed to helping boys through scouting and fundraising efforts. He emphasizes the importance of providing meaningful experiences for young people to keep them engaged and active.

"I am still fighting the appeal process; but this is not going to define who I am as a person. I have 'Rangered up.' Even though I cannot attend meetings and be in contact with scouts, I have found another way to support our local scout program.

"Near my home is a large piece of property our district has leased since the 50's from the local Air Force base. With approval of the local scoutmaster, who is a good friend of mine, I focus my efforts on improving the facility, and coordinating with folks on the base. We recently got a new gate installed and got the road graveled. I just keep working with the base to improve the facilities.

"I will continue to support my local troop. I buy things for them. I do things for them. I go down to the building, and I fix stuff. I do what I can for them, but I cannot officially be in that capacity. But I am not going to let that event stop me from being a Scout. And that is just all there is to it."

Ed does not know when his appeal will be resolved but he still has purpose and direction and is getting on with his life. Besides maintaining the local Boy Scout facility, he is spending quality time with his spouse of 27 years, Kimberly, their daughter, Shelby, and son Torsten. He is doing a frame off restoration of a '73 Ford Bronco. He also spends time hunting and fishing with his BFF, a chocolate Labrador Retriever named Samantha Rose!

* * *

Joe Edward Wishcamper was born and raised in Abilene, Texas. His father was a Navy veteran of World War II and the editor of the local newspaper. Joe excelled in school and accepted an appointment at the United States Military Academy in 1968. Joe said his inspiration to go to West Point was generated inadvertently by his mother because life with her was miserable. He wanted to get out the door and get away from his mother, and life in West Texas was not particularly pleasant.

Joe excelled at West Point and graduated in the top 15% of his class and could have gone to any branch in the Army, but he decided he wanted to be a military intelligence officer. In those days, noncombat arms lieutenants were detailed to a combat arms for a year so that they would better understand the front-line soldiers that they would support through the years.

His first assignment was to a mechanized infantry battalion in the 4th Infantry Division at Fort Carson, Colorado. In very short order, Joe was a mechanized infantry platoon leader, company executive officer, and then selected to be the support platoon leader when one of his classmates was relieved (fired). It is one of the most important lieutenant positions in a mechanized unit because you have responsibility for keeping the armor and armored personnel vehicles constantly fueled and maintained. Joe enjoyed the challenge of the job and being appreciated by the chain of command in the unit.

One day Joe got a call from a former professor at West Point who asked how things were going. Joe told him he really enjoyed the infantry. The professor then told him that he was in the wrong branch and that he should request a branch transfer to the infantry, which Joe did immediately. A couple of months later, Joe got word that his branch transfer was approved, and he was now an infantry officer. Shortly thereafter, Joe requested assignment to the 1st Ranger Battalion but was told he was too junior; so, he applied to the 2nd Ranger Battalion which was just being formed. Without an interview, but surely based on phone calls by LTC Bo Baker, the new 2nd Ranger battalion commander, Joe was accepted and was soon on his way to Fort Lewis.

Joe was one of the first lieutenants to report into the Battalion and got there in time to go to Fort Benning for the initial cadre training. It was there that he met his soon to be company commander, CPT Lawson Magruder, who was finishing up the Infantry Officer Advanced Course.

Joe said this about his and the other new Bravo Company officers' initial meeting with CPT Magruder, "Our introduction to him was whining, sniveling, belly aching, complaining, bitching, and moaning about what a bum experience we were having because we did not have a company commander. The other company commanders there were running all over us and screwing us out of resources. They were abusing us left, right, and sideways. So, we were not happy about it.

"I have often thought he probably came out of that meeting discouraged because we had very little positive to say."

But Lawson was anything but discouraged, because he was very impressed with the quality of the Bravo Company lieutenants. They had diverse backgrounds, were excited to be a part of something special, and ready to get to work. They were also not shy about expressing their opinions. Lawson left there feeling blessed with those selected to be a part of his team.

Joe would lead the 1st Platoon for the next 18 months until he was promoted to captain. He did a tremendous job overseeing squad and platoon training during the first six months leading up to company and battalion level training and excelled on every platoon, company, and battalion mission. He also participated in challenging, dangerous training in the subarctic, jungle, and desert environments.

Physically fit, highly intelligent, and tactically smart, Joe was totally committed to the development of his soldiers and focused on standards. Having never experienced an airborne unit, he was willing to learn from other paratroopers and his fellow lieutenants. He was also in awe of the quality of his NCOs.

Joe stated: "One of the things that I remember was standing in front of a formation of my platoon of Rangers and the thought that entered my mind was, *Joe, you have got to be worthy of these guys. You have to live up to their expectations, and their expectations are very high.*

"And so, I didn't have an opportunity to screw up because I had all these incredible NCOs behind me; squad leaders like Grzelka, Rekasis, and Chiotte."

Joe relates this story about the level of trust and confidence he had in the competence of his NCOs: "We had the final operation on the ten-day battalion certification exercise. I was the platoon leader that was first in order of march for the company. The daylight attack involved navigating through some of the densest, deepest, nastiest terrain that Fort Lewis had. And in retrospect, I think they probably chose the objective for that exercise based upon that.

"I mean, they said, 'Where is the nastiest terrain on Fort Lewis? And that's where we're going to send these guys.' And so, my platoon was in the lead, and this terrain was super thick, super tangled, and you certainly couldn't walk very far.

"And at some point, CPT Magruder called me back to where he was and asked, 'Who do you have on point?' I said, 'Grzelka and Rekasis. I put them up front because I have the greatest of confidence that those leaders could follow this azimuth to get where we are supposed to go.' And I certainly did not want to lead the platoon, and company astray.

"CPT Magruder then stated, 'The evaluator thinks we are off track or off azimuth.' And I responded, 'No way. I'm confident that we're on track.'

"CPT Magruder did not cross examine me in great depth and just said, 'let's go.' About three and a half hours later, thrashing and crashing through this incredibly difficult, tangled terrain, three guys stepped out onto the trail, Grzelka, Rekasis, and me. And were about five feet away from the objective at that point.

"Consequently, the company achieved outstanding ratings on the exercise and were deemed combat ready for worldwide deployment."

Besides his pride in the NCO's he had to help him build the 1st Platoon from scratch, Joe shared these reflections on his time in the 2nd Rangers: "The most prominent feeling I had was a feeling of responsibility that the Army and Lawson Magruder and to a much lesser extent, LTC Baker had given the responsibility to me for this platoon. And I needed to show myself worthy of that and how would I do that? And I guess number one was to provide a leadership profile every time there was an opportunity to do that. Which is to say, keep myself in shape, keep my unit together, keep the communication open with my NCOs, and leave the platoon combat ready when I leave it.

"So, everything that I did, everything that we did, thereafter, was to try to measure up to this responsibility that I felt was on my plate. And our training program gave me the opportunity to test that proposition at every turn. Everything that we did involved some kind of test that pushed us to the outer limits of that envelope. I felt the responsibility to meet those expectations myself and set a standard and an example for the platoon and the Rangers in it.

"My primary focus was on the NCOs, because this was all flowing downhill. If I set the example and performance for the NCOs, they would do the same thing for those under them. So, it was an intense exercise in meeting not only Lawson's expectations of me, but my own expectations of myself."

In June of 1976, Joe and several other lieutenants in the battalion were at the four-year mark since their commissioning and were promoted to captain. Unfortunately, the battalion had a limited number of captain company command and staff positions, so the majority had to leave the battalion. The majority went directly to company command positions in the 9th Infantry Division. Joe ended up as the headquarters company commander for the 2nd Brigade responsible for leadership of a company that consisted of primarily senior NCOs and officers on brigade staff. It was not a rifle company where he would have thrived.

Unfulfilled and quite honestly depressed after leaving 2nd Rangers, Joe made an impulsive decision to follow the recommendation of a West Point classmate and decided to leave the Army when his commitment ended to become a lawyer. He was quickly accepted at the University of Washington's School of Law.

Joe enjoyed his three years in law school because he learned more about decision making and how to analyze a problem. He felt that law school was one of the most positive experiences of his life and he was glad he did it. But the next 30 years practicing law were unfulfilling.

In Joe's words: "The moment I got out of law school, I wish I had somehow figured out a way to get back into the Army. And part of my disenchantment with the practice of law is that my experience was around dispute resolution. If you want to see the worst side of people, get involved in a dispute, get involved in a situation where you have an absolute duty to perform your very best to try to help this person in a dispute. Well, the problem was the dispute more often than not involved something that the person did or did not do that they would not take responsibility for. And they were not going to take responsibility for it because, in their mind, it wasn't their damn fault and because it wasn't their fault, they didn't want to take any responsibility for it. They didn't want to take any liability, they didn't want to pay any money, they didn't want to pay me, that's for sure.

"And so, it was an experience that brought out the worst in people, not the best people who are in a dispute. Human nature comes forth in a roaring fashion, and human nature is that people do not take responsibility for what they do or say, and they do not want to take responsibility because if they ever take any responsibility for their actions, it is the same thing as taking blame. And they do not want to take blame because people who take blame do not win.

"I was seeing how my fellow human beings behaved in circumstances where they had screwed up and they were not taking responsibility, never would. And they were looking for other people to blame. And among the people to blame for sure is the opposing counsel, the attorney who was representing the other guy. And so, if you are in a position where circumstances are bringing out the worst of people, it is hard to keep a positive attitude."

Joe's experience as a lawyer was diametrically different than his experience as a Ranger platoon leader where he accepted personal responsibility, responsibility for his Rangers, and admitted mistakes. He had to endure a profession that was often contrary to his personal values. But Joe "Rangered up" and through sheer grit and determination and support of his beloved spouse Susan, endured and succeeded in his chosen career.

Joe's legal profession may have provided a nice income, but it left him unfulfilled and lacking a clear purpose in his life. But as many of us do, he filled that void when he was introduced to mountain climbing by a fellow lawyer. With Mount Rainier nearby, Joe threw himself into learning complex climbing techniques. He joined a local mountaineering club and after six months gained his basic climbing certification and soon was climbing Mount Rainier. He was extremely physically fit and loved the challenge climbing provided. He soon enrolled into the two-year advanced climbing course.

To graduate from the advanced climbing course, students had to go on an expedition somewhere. Undaunted, Joe climbed with his team the highest mountain at 22,837 feet above sea level in the Western and Southern hemispheres: Mount Aconcagua in Argentina.

As Joe described the incredible experience: "I will always remember how difficult it was once we got above 20,000 feet. There was not enough oxygen. We were in absolute top physical condition but if you don't have oxygen, you're not going to do well. And I recall telling myself and anybody else who would listen, I now know why expeditions to climb Himalayan peaks typically go into oxygen at around 20,000 feet. And the reason they do is because there isn't any oxygen above 20,000 feet. It's tough. The last 1,000 feet on our climb to the summit of Mount Aconcagua, we would take two or three steps and then we were panting, panting, and panting, and we just could not suck in enough oxygen to catch our breath.

"So, it was an unpleasant grind, but we made it to the top." Once again, in true Ranger fashion, Joe accomplished the mission and gained his advanced climber certification. Joe sunk himself into climbing for the next 15 years. It was not only challenging but was a distraction from his legal profession.

One of his most rewarding climbing experiences occurred on a climb up Mt Rainier. "We had some of the best mountaineers in the Pacific Northwest, including a fellow by the name of Jim Whitaker, who was a legend. He was the second American that summited Mt. Everest. Whitaker and other mountaineers in Washington State organized a group of blind people to go climb Mount Rainier. We provided the guides and the cadre to take them up.

"For blind climbers, let me give you some perspective. It is an absolute physical, mental, gut bust to climb Mount Rainier. It happens all the time, but it is not easy. And there's sort of an analogy between climbing Mount Rainier and going through Ranger school for people who are blind and cannot see what they are doing. I was privileged to be brought into this group to serve as a guide. One of the people that was in this group of blind folks was a lady. She just ran out of gas about two thirds of the way up, and she could not go any further.

"And so there was a lot of discussion about what to do with her. We were not going to leave her alone. That could involve potentially life-threatening risks. So, if we are not going to leave her alone, who is going to stay with her? And for reasons I did not know at the time and still do not know, the other cadre members chose me to stay with this gal while the rest of the team trudged on to the summit. And so, I stayed with her, comforted her, and gave her encouragement and kept her from flying off the handle.

"And that was a significant thing from her perspective, that I had been chosen to be her guardian angel, if you will. It was quite an honor." It is not surprising that Joe Wishcamper was selected to take care of the lady in distress. He knew how to calmly and passionately take care of her just as he had done as a young Ranger platoon leader.

Joe had to give up mountain climbing several years ago when he was diagnosed with congestive heart failure. It was becoming harder for him to just breathe naturally. Just climbing stairs has become a challenge for him. He has undergone major cardiac procedures and surgeries over the past few years and with typical Ranger toughness is fighting the battle to return to a good purpose filled quality of life. He is being cared for by the absolute best cardiology team in Seattle and his "battle buddy" for the past 41 years, Susan.

* * *

We would like to close this chapter with some thoughts from Joe on why his time in the Army was so meaningful to him: "While in the military, you can meet and work with some of the most outstanding people you will ever know. People like Lawson Magruder, Fred Kleibacker, Frank Rekasis, Tim Grzelka and Jim McNeme. You will never meet and work with those quality folks until you are in the military. And even though I had a relatively short career, I can tell you that even to this day, the people that I met in the military were closer to me than my own family members.

"And I have a buddy who retired from the Air Force after 22 years. He graduated from the Air Force Academy. He and I have talked about this, that he and I are closer than his own family members. He is closer to me than his own brother. So, there is a bond between people who served in the military, but not among those who have not served. That bond is priceless and will affect you for the rest of your life. I can spend ten minutes with Lawson Magruder and feel like I got ten years' worth of joy and experience out of it."

Joe closed with: "The relationships that you have in the military, they will be much more intense than relationships that are outside the military, and they will endure longer than relationships outside the military. The fact that we have Bravo Company reunions is testimony to the fact that those relationships retain their vitality over the years. They really do. So, if somebody is contemplating going into the military and they ask themselves the question, "Why should I do this?" My response is, "You are going to form relationships and friendships that are going to endure the test of time, the test of distance."

14. BECOMING A GOOD CITIZEN

Once you have served your nation, you will truly understand the meaning
of citizenship: the Constitution because you protected it, the rights of
all citizens, and the need to abide by our laws. You will be inspired to
continue to serve others.
—CPT Lawson Magruder

From left: Kim Maxin, Shaun Driscoll, Mark Wheeler

THE PHONE RANG, "HELLO."

"Hey Kim, you got a minute. We need your help."

"Sure, what do you need?"

"We have a client who recently lost his wife. He's 75, a former lawyer,
and a former Army reservist. Since his wife died, he kind of shut himself
in. We were hoping since he's a Yankees fan like you, you two might hit
it off. Plus, both of you guys have ties to the Army. And your experience

in counseling and volunteer work with Home Instead and Alzheimer's patients could be beneficial. Can you help us out?"

"Absolutely, what's his name, address, and phone number? When do the other aides typically go over?"

"I'll email his info. Someone is there every day at 9am for three hours. Can you do it tomorrow?

"I'll be there," Kim replied.

Kim arrived a little early and right at 9am rang the doorbell. Chris opened the door and greeted Kim. He slowly looked Kim up and down. Kim was wearing a beige baseball cap with a small 2/75 Ranger Scroll over the Ranger unit crest embroidered on the front of the cap. He then looked at the Yankees Jersey Kim was wearing.

Kim reached out to shake the man's hand and said, "Hi, my name is Kim Maxin. You must be Chris."

Without hesitation he replied in a soft deadpan, "That's right. So, you're a Yankees fan?"

"Yes sir!" Kim enthusiastically replied.

"What's that on your hat?" he inquired.

"That's the unit I served with in the Army. I was an Airborne Ranger. I served in the 2nd Ranger Battalion at Ft. Lewis. I'll tell you all about it if you like. Why don't we sit down. Do you have a favorite place to sit?" Kim responded and asked.

They would go on to talk for the next two and half hours. Kim telling him all his best Ranger stories and listening to Chris talk about what he did in the Army. Then they talked about the Yankees. After the visit, Kim called the office.

"Hey, I thought you said this guy didn't like to talk? I don't know who told you that, but I think he is fine. He just misses his wife."

Being a good citizen is about the sincere obligation to take care of others whether it is convenient or inconvenient. It's about always putting others first, "paying it forward" every day. In this chapter we are going to highlight three Rangers who went on to become exemplary examples of good citizens to thousands of kids and elderly folks over the course of their post military careers, Rangers Kim Maxin, Shawn Driscoll, and Mark Wheeler.

Kim Arthur Maxin was born and raised in the north country of upstate New York. That area of the country was full of former WWII vets and patriotic sentiment. But like most male teens during the late 60's, Kim's upbringing was all American, chasing girls, drinking beer, playing baseball and soccer, but most importantly in upstate New York, it was all about playing the game they loved more than anything else – hockey. The long, dark and cold winters didn't leave much to do after school or on the weekends, so everyone played hockey or rode snowmobiles.

In his senior year of high school, Kim applied and was accepted to attend college, getting his deferment from the draft and most importantly, to play college hockey – the real reason he wanted to go to college. After he graduated high school in 1970, he started college that fall. He completed his first semester, and decided to continue for another semester since he didn't have much else to do. He went to the registrar's office on a deep, dark winter day. It was minus ten degrees with a biting wind chill. The line was as long as a football field. There was no way he was going to stand in line for hours in subzero weather to register for some dumb classes. There had to be something else.

Kim explained, "I knew my draft number was low. It was a seven, so I was going to get drafted if I quit college. I decided the heck with it, I'll just enlist. I went down to the army recruiting station and remarkably they were closed, in the middle of a war, because apparently, they had all the draftees they needed. So, I just let them draft me."

Kim entered the Army in April 1971 and was selected to attend Advanced Infantry Training (AIT) after Basic. He had no specific assignment other than "the needs of the Army." During the Vietnam War, draftees were typically given a two-year contract, since most would be going to Vietnam. The first six months were training: basic and AIT. Then they were off to Vietnam for a year. If they survived, they returned to the States to be processed out of the service at the completion of their second year.

Kim described what happened next, "So about halfway through Basic, we're sitting there getting the normal lectures all trainees get. How to spend your money, how not to spend your money, like don't spend it all at the strip bar because you're only making $98 a month. And then we get the one about the diseases you can get from all the "off limits" joints, like massage parlors downtown. And, of course, all the myriad other briefings about what 'you can't do' for about three hours. We're all trying not to fall asleep cause we didn't want to be doing pushups and getting yelled at by the Drill Instructors. The last guy finished and suddenly this NCO from the 82nd Airborne stomps up onto the stage. He looks like a giant in his dress greens. He's got these spit shined jump boots, his greens trousers are neatly tucked into the tops of his boots, he has a chest full of medals, and a cocked garrison cap with a glider patch on the left side of his cap. He puts his hands on his hips and loudly announces, 'Listen to me, privates! There are only four kinds of people in this army! The sick, the lame, the lazy, and the Airborne Crazy!'

"Well, he's got my attention now, so I'm listening. And he says, 'If you want to jump out of an airplane,' which I'd never even ridden in an airplane in my life, 'you're going to get an extra $55 more a month. Now I'm only making $98. I think to myself – this is a no brainer. So, I just said, all right. I had to reenlist a couple of days later to be able to go airborne and then I volunteered for Special Forces. I figured, if I must go to Vietnam, go with the best. I finished basic and went to Ft. Lewis, WA, for my Infantry AIT. When we completed our AIT training, the entire training company got orders for Vietnam except for eight of us, the only ones who had volunteered for Airborne School."

After finishing Airborne school, Kim was selected to attend the Special Forces Qualification Course at Ft. Bragg, NC (today Ft. Liberty). While he was in the holding company waiting to start the course, he had a falling out with one of the NCO cadre members who ended up putting him on guard duty as punishment for the foreseeable future.

Kim decided this was not what he wanted to do and volunteered for Vietnam. A few days later, when he went to pick up his orders, he was told they had been changed. He was going to B Company Rangers, 75th Infantry at Ft. Carson, CO. Kim had no idea why they sent him to the Ranger Company, but he wasn't complaining. The only problem was, he had no clue what a Ranger Company was.

Kim arrived at B Company Rangers a few days later, only to find himself one of only a handful of privates in the company. Most of the soldiers were returning combat veteran NCOs from both Special Forces and other Ranger Companies in Vietnam. Most were Staff Sergeants (E6) or Sergeants (E5). B Company was a strategic reserve company for the 7th Corps focused on defending Europe against the Soviet Union. It was never deployed to Vietnam as a unit during the entire war. Individual Rangers would go and fill in at other Ranger Company's in Vietnam and then return to B Company.

Although just newly married, Kim was sent to three different schools over the next six months where he would learn to be a parachute rigger, successfully completed Ranger School, and then attended NCO schools. Afterwards, he was promoted to Sergeant (E5) in 1972.

The company moved to Ft. Lewis, WA, in 1973 and was disbanded in 1974 when the newly activated 2nd Ranger Battalion was formed at Ft. Lewis. Kim volunteered for the battalion and arrived in early 1975 as a Sergeant (E5) and was assigned to Bravo Company, 3rd Platoon as the 3rd Squad Leader. Kim's new platoon leader was his old Executive Officer in old B company, LT Eldon Bargewell. He was in good company since his buddies Al Kovacik and Tom Gould, also from the old B Company, were assigned to 3rd Platoon as well. In short order, Kim was promoted to Staff Sergeant (E6) and would stay as the Squad Leader of 3rd Squad until he departed. Kim recognized he was part of something special from the very beginning.

He said of his time in the Rangers, "One of the most important things I learned there was always being prepared, just totally being prepared. Everybody always says you can give more than 100%. There's no such thing as more than 100%. What it is, it's giving it 100%, or your very best, every single day without fail. When I was at work, I worked to become a better leader and to make my people under me better soldiers. I did what I was expected to do as a leader and to always represent the battalion honorably, both on and off duty. It was about building character, bringing honor to the organization and yourself. It was doing our sworn duty and keeping our oath to serve our nation to the best of our ability. It was our duty."

Kim would serve honorably as 3rd Squad Leader for two and a half years, departing his beloved Rangers in 1978. He decided to take an assignment in Hawaii and was assigned to the 2nd Brigade, 25th Infantry Division, on the island of Oahu. It was downhill from the beginning.

Kim described the culture, "The 25th was just a mess, especially after what we did in the Rangers. There was no leadership. I was seeing captains with combat experience and valor awards on their chests leaving because they just couldn't deal with it. The battalion commander was relieved, there was no physical training, no communication between what we were going to do out in the field and the operations shop. No Operation Orders, nothing. They would just send companies into the field and tell them to camp out for a week, telling them, 'Just go to this point and stay there. We'll get you later.' I mean, it was just poorly done.

"And then the all the drug issues became a big issue. They just had a lot of drug issues at that point in time. And I just didn't like being in a leg (non-Airborne) unit. I had so much more to offer. So, when my enlistment ended, I decided to get out and go back to college." Kim had spent eight years in the Army, seven of those with Ranger units, when he decided to get out and return to college.

Kim would return to his roots in upstate NY and go to Potsdam College, receiving his bachelor's degree. He would continue to Colgate University to earn his master's degree in counseling. He would spend the next

30 years as a school counselor in rural high schools and with the Board of Cooperative Education Services (BOCES), which focused on special education, adult education, and GED programs. He spent his last 12 years as a BOCES counselor at the Technical Education, Power Mechanic, and Auto Technician Vocational School serving nine different high schools in the area.

Kim chose this career as an opportunity to help young people as a service to his community. During those 30 years, he also served as a school board member, president of a local chapter of school counselors, and as president of a sports booster organization which he helped establish. Kim also said he went into counseling because of the challenge it presented, since it depended on who was sitting in the chair across from him during counseling. He described the challenge as, "You didn't know if it was going to be a parent walking in, a teacher that was having a problem, or a student needing advice. Every day was a challenge, an extra challenge. And I love challenges, and I loved that the Battalion challenged me, and I think I met that challenge pretty well."

Kim became one of the most trusted members of the faculty in every school he worked. He spoke about his years as a counselor, "I liked to be in rural schools, so I knew everybody. I usually lived in the village so I could get to know the families. When the kids would come back from college, they'd come back to the school, and the first person they looked up would be me or other select teachers that had made a difference in their life. They weren't looking for my approval, they just wanted to thank me. Here's an example. We had lost a student one night who overdosed on drugs. It was 4:30am. The other students didn't call the principal. They didn't call teachers or parents. They called me first because they knew I knew what to do. And I had a calming effect on them, just like our leaders in the Rangers had a calming effect on us."

When asked what his purpose was in life, Kim responded, "Kindness is the highest level of wisdom, and I'm always reaching out for that wisdom. I am kind to everybody until they don't deserve the kindness. But I'll always reach out. I try to do it in the neighborhoods first, start with my neighborhood and the people around me, and then spread that out so I don't run all over the country doing it. I just do it where I am.

"I start with individuals and go out in concentric circles. I start at my own feet and work out. Just like we did in the Rangers. I'm doing what I did as a squad leader, at the squad level. I started with the individuals, the teams, then the squad."

Kim's journey from an Army draftee to an Airborne Ranger squad leader to a rural school counselor is an exquisite example of a good citizen. He took what he learned from being challenged in the Ranger Battalion, growing in his self-confidence and forging his commitment to his fellow human beings. He combined knowledge, hard work, and respect to build lasting relationships. First as a trainer and mentor to his troops as their squad leader in the Rangers, and then he brought those skills forward into being a counselor who listened carefully to help people solve their problems. It is the most noble of professions.

After retiring from counseling in 2013, Kim turned to helping the elderly, bringing his years of experience to give back to his community, which brings us back to the beginning of his story with Chris. Kindness, selfless service, sense of duty, personal honor, and an unwavering commitment to positive action… these are the values of a good citizen we all should seek to emulate.

One of Kim's very favorite people was Richard. He was a 98-year-old WWII army soldier and recipient of the Distinguished Flying Cross. After using his GI Bill at Syracuse University, he raised a wonderful family while being a physical education teacher, coach for swimming, golf, soccer, and football. Richard passed away recently and was truly a very extraordinary man to Kim because he was always very humble.

***

Guiding young people to be successful and believe in themselves when they lack self-confidence or perhaps have special educational needs takes a special person. It is especially hard in environments where the culture is prejudicial towards those attempting to positively influence young students about America due to their own confirmation biases. Building trust and respect with both the students and the school administration is paramount and takes time, patience, and demonstrated success.

Such is the story of Ranger Shawn Driscoll.

Today was the day for the new freshman, cadet David, to do his daily brief. It would be in front of the whole class, on a news article he had picked. The assignment was that he had to read the article he picked, write something about the article, and brief it to the entire class. Today was his first time briefing his class and he was terribly nervous. He had never in his young life done anything like this before. He was a shy, foster kid who had low self-esteem and confidence since he believed nobody wanted him. He had been moved between foster families for most of his life. And once again, he had a new family, was at a new school, starting over – again.

Chief Warrant Officer 4 Shawn Driscoll, senior officer for the Spaulding Union High School JROTC program in Barre, VT, had successfully encouraged David to join the program when he had arrived on his first day of school. Because David was new and had an obvious lack of self-confidence, Chief Driscoll had assigned him an upper classman to be his 'battle buddy' and mentor him. In this case, it was the Cadet Corps Commander, a talented senior named Aiden.

"David, you're up," Chief Driscoll announced softly.

David got up and shuffled to the front of the class. He was nervously holding his papers and looking down. He got to the front of the class. His mouth was bone dry, and his heart was racing and seemed like it was trying to jump out of his chest. He was scared to death and petrified to look at his classmates, afraid they would be making fun of him and mocking him.

He took a deep breath and looked up. He was surprised to see all the students, seated and ramrod straight at their desks, were looking back at him. No one was mocking him. But many were smiling, and Aiden was giving him a big thumbs up with a huge grin on his face.

"Ok David, let's get started, son. We have 20 other cadets waiting. You'll do fine," the Chief encouraged.

David cleared his throat and gave his briefing. It was all over in 45 seconds.

"Good job, Cadet! You can take your seat. Cadet Johnson, you're up," announced the Chief. And so, it went for the next 20 minutes. Each cadet, standing in front of the class, briefing the other cadets on their article until everyone had successfully accomplished the assigned task.

Shawn Patrick Driscoll was from Boston, MA. His dad had been a Marine and eventually transferred to the Air Force and retired there. In his sophomore year of high school, Shawn was intrigued by serving in the military and checked out the Marines and the Army, where he found out about the Rangers. As luck would have it, his mom had a cousin, Lee, who was an instructor at the Army Ranger School at Ft. Benning, GA (Today Ft. Moore). Shawn was introduced to Lee and visited with him that Christmas at Ft. Benning. They hit it off. Shawn spent the following summer with Lee and his wife at Ft. Benning, going to work with him every day, watching new Ranger students go through their crucible.

Lee was reassigned to the Ranger Camp at Eglin, AFB near Ft. Walton Beach, Florida. He suggested Shawn should come down and spend his junior year of high school with Lee and his wife at the Ranger camp to experience the Ranger culture to see if he liked it. Shawn's parents met with Lee and agreed. Shawn moved to Florida to stay with Lee and his wife and experienced firsthand what it was like to live the Ranger lifestyle. Shawn sums it up, "I just lived the Ranger culture as a young guy and loved it. Lee and that experience, were my biggest influences in enlisting for the Ranger Enlisted Option and the 2nd Ranger Battalion."

Shawn officially entered the army in June 1975 at 17 years old. He went through the normal pipeline of basic, infantry AIT and Airborne School, arriving at B Company in October 1975. He was assigned to 3rd

Platoon, Weapons Squad as part of a three-man, machine gun crew, as an ammunition bearer. The battalion was experiencing a high personnel attrition rate, due mostly to the intense mental and physical training. So, most squads in the Battalion were under strength. Shawn was about to be thrown into his own personal crucible right out of the gate.

Training was ramping up by October, so Shawn had a lot of learning to do. He'd learned how to disassemble and assemble the M60, general purpose, medium machine gun that the Army used at the time and had fired it once during Basic Training for familiarization, but that was it.

Shawn remembered, "My squad leader was SGT Tico Layton, a great squad leader. He taught me everything about the M60 machine gun, crew drill, and everything about every piece of equipment that we had to carry. I was assigned as an 'ammo bearer,' but didn't realize that, because of the personnel shortages, I was also the acting assistant gunner. So, instead of a three-man gun crew, we were a two-man gun crew. So as the assistant gunner, I carried the gear that weighed as much as what a gunner carried. At the time I weighed 130 pounds soaking wet. I carried at least that weight in gear, sometimes maybe more."

Shawn went to Ranger School about a year later and would stay in 3rd Platoon weapons squad as the squad leader for another year. His last position in the platoon was as the 2nd squad leader before reenlisting for Special Forces.

When asked what he thought about the high attrition rate, Shawn replied, "When I got there, we had a lot of people leaving over the first couple years, they just couldn't cut it. They would be sent down to leg land (9th ID). I'd hear guys in the barracks saying, 'Did you hear so and so wimped out?' That word scared me. I didn't want to 'wimp out.' Being a weapons squad guy, it's like, I got to carry all this weight, but we have an enormous responsibility to the platoon. Those three machine guns were the most powerful weapons we had in the platoon. We protected the platoon.

"My attitude was, I will die on my feet, but I won't quit. There's no way they're going to say, 'Driscoll wimped out.' So, what I witnessed and immediately learned from the guys who had no intention of quitting was their intestinal fortitude, their loyalty to their squad, the platoon, the company, and the battalion. It was a force."

Shawn reenlisted for Special Forces training in October 1978 and for the next 20 years served in various Special Forces assignments with 10th Special Forces Group, both as a senior NCO and Warrant Officer. He deployed in Operations Desert Shield, Desert Storm, Provide Comfort (Iraq), and Silver Anvil (a non-combatant evacuation operation in Sierra Leonne, West Africa). He served as an Operational Detachment Alpha (ODA) executive officer and team leader for 10 years. He went on to manage the SF Warrant Officer Program for three years and was selected to command the SF Warrant Officers Basic and Advanced Courses. This is where SF Warrant Officers learned to do their jobs. His 23 years in elite units with superior leaders had prepared Shawn well for his next career.

Shawn decided to retire in 1998 as a Chief Warrant Officer 4 after 23 years in the Army. It was time. He had three kids, all under seven, and it was time to go to Vermont, build their dream home, and put his kids and wife first. He wanted them to grow up normally, surrounded by their cousins, aunts, and uncles. While he was building their family house, he applied for and was accepted into the Junior Reserve Officers Training Corps (JROTC) program. He would eventually work at two different high schools in Vermont. The first was a school in northern Vermont near the Canadian border and then five years later he transferred to the high school near where he lived and where he would remain until retiring 14 years later.

Shawn was proud of everything he had done in the Army. But working with kids to help them navigate the crazy cultural stuff and teaching them the same values he had brought along with him since his early Ranger years was not only personally gratifying, but probably the most important job he had in his entire career. He touched hundreds of kids over his 19 years in JROTC. Today, he is quietly retired with his wife near Ft. Walton Beach, FL, very near where his pivotal junior high school year would change his life forever. He and his wife raised four kids, all of whom served in the military, with one currently still serving.

He remarked about his career, "I carried with me every lesson I learned throughout my military career. All those early leadership models I was exposed to in the Rangers were my foundation that have stayed with me to this very day. They are indelibly implanted in my brain. I carried and used all of them as a JROTC instructor."

He continued, "Honestly, I viewed JROTC almost like an unconventional warfare mission. Here's two guys, an Officer and an NCO, they parachute into a rural high school and are tasked with what would seem an impossible mission. To do this kind of an incredible thing – transform kids who come from all walks of life and socio-economic backgrounds into good citizens and future good leaders. Our job was not to make soldiers, our job was to help the kids understand how special America is, learn about the Constitution, and to be good citizens.

"If they wanted to serve their country in the military, we helped to facilitate that, but it was not our primary mission. If they wanted to enlist, we could get them a promotion to E3 or E4 because of attendance in JROTC and their academic achievement. We focused on helping them with their academic achievement. If they wanted to go to college, we could help them apply for a ROTC scholarship, sometimes a full ride."

Shawn went on, "Here's one for you. A friend of mine from 10th Special Forces Group messaged me recently on Facebook. He posted a picture of himself with another officer and asked me, 'Hey Shawn, do you know this guy?' I recognized him right away, it was David, the shy foster kid who did four years in my JROTC program. He was now a Major in the Army. I was never prouder."

Kim and Shawn transformed individuals and their communities. All with a simple message, as citizens they have a humble duty to be role models and to effect change in their small world. It's an incredible responsibility, a noble cause, unpretentious, innate, but inspired through their Ranger experience. Such is the story of Ranger Mark Wheeler.

* * *

The whistle was shrill, "Everybody on me now!" barked Coach Wheeler.

As the Bonney Lake High School boy's baseball C-team gathered around their coach, he ordered them conversationally, "Okay, everybody get down and give me 20."

Mark dropped to the ground before the team could get on the ground and began to knock out 20 perfect Ranger pushups. The boys around him did the same. Mark jumped up and waited for them to finish. As they finished, they got up one by one up from the ground and focused on their coach.

Somebody screwed up they thought. Coach doesn't make us do push-ups for fun. The team had been divided up to work on their own individual skills like batting, pitching, fielding, and other core baseball skills. They knew the coaches couldn't watch everyone all the time, so given human nature, some of the kids would goof off instead of practicing.

In the center of the group, Mark held their attention with silence as he looked around the squad of ballplayers, quietly eyeing each player, then smiled and said, "Gentlemen, integrity and character are the two main things that you will carry with you for the rest of your life. It's a choice. If you think you can get away with something like skipping an assigned drill or not completing an assigned task because you think no one is watching you, you lack integrity.

"Integrity is doing what you are responsible for doing with 100% effort when no one is looking. If you lack integrity, you lack character. Character is about your individual moral strength. If you lack character, no one will trust you. If the team doesn't trust you, we can't win games. Does everyone understand me?"

"Yes, coach," they replied sheepishly in unison.

"Alright, get back at it," he ordered.

Mark Thomas Wheeler grew up during the Vietnam War, watching it unfold on the evening news. He and his friends would play Army on the streets of Pittsburgh, PA after school and on weekends. His dad was a Korean War veteran and instilled in Mark a love of Country that has stayed with him to this day. Just before his junior year in high school, his family moved to the small rural community of Linesville in northwest Pennsylvania near Erie, where his dad grew up.

Like many of his fellow Bravo Rangers, Mark was influenced by the positive patriotic TV programs and movies he watched on TV during his youth. Movies like John Wayne's "The Green Berets" and TV series like "Combat" and "The Desert Rats," influenced him about wanting to be a 'commando.' When Mark was in high school, he had a job where many of the employees were veterans who had served in World War II, the Korean War, and the Vietnam War. He would listen to their stories, and they instilled in him a strong desire to serve his country.

Combined with his sense of desire to serve, he was also pragmatic, he decided that the military was a way to pay for a college education, serve as others had that he respected, get out of small-town America, and maybe see some of the world.

In his senior year, he went down to the Army recruiting station and asked to join the Special Forces (SF) to pursue his dreams of becoming a commando. His recruiter told him the only way he could get to SF was by joining the Rangers first. This was in December 1975, just a little over a year after the 2nd Battalion had been activated and around the same time it was completing their Combat Certification ARTEP. This first year had been rough on retention and both the 1st and 2nd Ranger Battalions needed solid Ranger prospects. The Army was offering substantial signing bonuses to qualified recruits. In Mark's case, that was a whopping $2,500 to join the Rangers. To put that into perspective, that is around $14,600 in 2024 dollars. That sealed the deal for Mark.

He was 17½ years old in his senior year at high school. Mark enlisted in the Delayed Entry Program under a Ranger Enlisted Option contract for the 2nd Ranger Battalion. He wanted as far away from rural Pennsylvania as he could get. In June 1976, less than a month after his 18th birthday, he was on his way to Ft Dix, New Jersey for Basic Training. Mark completed basic and departed for Ft. Benning, Georgia (now Ft. Moore) to attend Advanced Infantry Training and Airborne school. He arrived at B Company in November 1976 and was assigned to 2nd Platoon as a rifleman.

Mark and the new crop of Rangers were getting to the battalion about the time it began to modernize and add mission sets to its prestigious light infantry resume. This was the nascent beginnings of what would become the U.S. Army Special Operations Forces and the Ranger Regiment of today. The focus was on adding new mission sets like airfield seizure, hostage rescue, enhanced pistol and rifle marksmanship, and long-range reconnaissance.

The battalion began to acquire new equipment including more vehicles, focus on expanding advanced skill sets like Military Freefall (High Altitude Low Opening or HALO), Underwater Operations Course or Combat Diver, Special Operations Training (SOT), Sniper school, Pathfinder school, advanced medical training (Special Forces Medics Training), advanced communications (Special Forces Communications Training), advanced demolitions training (Special Forces Demolitions Training), strategic target analysis and many other budding advanced special operations skill sets, techniques, and tactics.

It was an exciting time to be in the Battalion. Mark would spend 39 months acquiring many of the skills listed above. He stayed in 1st Squad, graduating from Ranger School, and becoming a team leader. In 1978, he attended the Special Operations Training Course and graduated as the Top Sniper. He attended HALO school shortly thereafter and was moved to Weapons Squad as the squad leader. Weapons squads were expanding from being just a machine gun support element. They were slowly experimenting with missions that began to resemble what the Rangers call today, Recce Teams. The squad would slip away in the middle of the night, no one knowing where they were going or what they were doing.

Missions were becoming compartmented; tactics, techniques, and procedures were becoming more sophisticated. These were the humble beginnings of today's post 9-11 Ranger Regiment. The Battalion was becoming a more elite special operations organization organically, setting the stage for what none of us could have imagined during those early days, one of four elite fighting forces in the U.S. Army Special Operations Command to be known as the 75th Ranger Regiment. They are the most deployed unit in the U.S. Army today.

Mark emulated his mentors. His squad leaders, SGT Joe Picanco and SGT Mitch Erickson, his platoon sergeant, SSG Robert Demoisey, and the Ranger he would replace as weapons squad leader, one of two original Bravo Company Rangers who made the ultimate sacrifice in 1993 during the battle of Mogadishu, SGT Tim "Griz" Martin. All these Rangers are highlighted in this book. They were part of the original 2nd Platoon plank holders who had paved the way and were now moving on to pursue new adventures in the Army and were handing the reins over to the next generation of Rangers, like Mark.

Mark spent the next years living his dream of being a commando. Working with FBI hostage rescue teams, supporting missions for the nascent Delta Force, conducting unilateral special operations missions all over the world. In the late summer of 1979, Mark did what all Rangers eventually do, he fell in love. The romance budded quickly, and he and his now wife of 44 years Paula, decided to get married. He had a choice, stay in the Rangers, get out of the Army, or see if he could find a job in the Army that would allow him to be home a little more often with his new wife to be. The current First Sergeant of Bravo Company, 1SG Voyles, affectionally known as Ranger Voyles or RV for short, was a legendary Ranger and heard of Mark's dilemma.

As Mark describes it, "1SG Voyles caught wind of it and pulled me aside. He knew I was thinking of re-enlisting and getting married. His advice was to go to "leg land" and see what the rest of the Army was really like. "After all," he said, "you can't hide in the Battalion for your whole career." Sooner or later, I would have to go to the regular Army to get promoted. So, in February 1980, I signed out of Battalion and signed into B Co. 3d Battalion 39th Infantry, 9th Infantry Division."

Mark arrived at his new company and had his initial interview with the company commander, a former 2/75 platoon leader. He asked if he could become his Training NCO so he could get some time with his new wife after three plus years at the tip of the spear. He was led on by his new commander to believe he wouldn't have to go to the field every time the company went for training. After two months, Mark realized that was not the case. He was in the field every other week and when they returned, he was not allowed to go home until the entire company's equipment had been cleaned, inspected, and put away, even though he had no troop leading responsibilities.

Disappointed that he was misled, Mark decided that if this is what the big Army was like, he wanted no part of it. However, he told us he learned a valuable lesson, "I learned there are two sides to every transaction, the SALES side, or promises made, and REALITY after you've bought the product. This is a lesson that would stick with me forever."

Mark departed active duty on Terminal Leave on May 21, 1980 – his 22nd birthday. He left as he had come in. He and Paula settled into the small community of Puyallup, Washington just outside Ft. Lewis. He enrolled in Bates Vocational Technical School in Tacoma as an Electrical Engineering Technician student. It was a self-paced, two-year program that taught the technical side of Electrical Engineering with all the same electives as a four-year college. Mark finished in six months. During this time, he joined the 12th Special Forces Active Reserve because he still wanted to serve. He spent four years on an A-Detachment before electing to go into the inactive Reserves for another four years.

In February 1981, after successfully graduating, Mark walked in off the street to an engineering firm and was hired the same day. He started as a draftsman and worked his way up to a designer position within one year. For the next three years, he OJT'd to learn the mechanical side of engineering and in October 1984 he went to work for a mechanical contractor as a design engineer. For the next four years he learned the construction side of the industry and eventually moved into project management, estimating, and sales.

For the following few years, he transitioned to being self-employed and in April 1993 started his own company as a side business. In August 1997, he left that company and officially went full time with his own company.

Today at age 66, Mark is still running his company managing major mechanical, electrical, and plumbing installations for large commercial projects. He works less on the design side of the business now and more on the commissioning side to ensure installation was completed to code and testing all the systems for proper functionality.

Mark has also been deeply involved with his local community for the past 40+ years. This is where his real passion lies, working with kids. He has spent over 11 years working with both Cub and Boy Scouts, 12 years coaching little league, middle, and high school baseball and 15 years in parish leadership in his church.

The single greatest quality that makes a good citizen is when one decides to make someone else's wellbeing their personal responsibility. That is Mark's story. When we asked Mark what he brought with him from his time in the Rangers, here's what he said...

"First, integrity. Whether I was designing or overseeing the construction of a building system, coaching, being a scout leader, or teaching Sunday School, I always tried to pass on and live by the thought of 'doing the right thing, no matter the cost.' I can't honestly say that I have never stumbled, but I like to believe I'm on the plus side.

"In the construction industry, doing the right thing isn't always the most profitable way to build the company's bottom line. Many corners can be cut that will never be seen once the walls or ceilings are finished. Even after starting my own company as a consultant, I knew there were times that I had no real say in what the client would follow through on. Not all believed in the same philosophy as mine. As a result, I have lost business. However, there have been some clients that have returned because of my values.

"Second, character. 'Who you are when no one is watching' is a key value. However, the second half of that saying is usually missed. Working with 15 – 30 baseball players at the junior high or high school level, you don't always have 'eyes on' every player. I always stressed to them that even if you think 'no one is watching,' you are watching yourself. You know if you're giving it 100% doing the drill. In the long run, that's more important than the coach watching you."

When we asked, "How would you advise leaders to lead today," Mark responded, "Three things. First, LEAD BY EXAMPLE. I grew up with a dad who had the 'Do as I say, Not as I do' philosophy. The Battalion taught me to SET THE EXAMPLE and be someone worth following. This is one of the core values that I tried to teach my kids and all those I've encountered since my days in Bravo Company. As an employer, or even as Project Manager in construction, I would not leave subordinates to work overtime by themselves. If a project needed to get finished, I stayed to work with them and get it done.

"Second, COMPLETE THE MISSION. In my dealings in the business world, having a reputation of 'being the one who gets things done' has paid dividends well beyond what I ever imagined. So many times, deadlines are set, and promises are made that never really mean anything, especially in construction. It's easy to not finish your task because someone else hasn't done their part. Don't succumb. Finish the task.

"Third, ALWAYS HAVE A PLAN 'B.' That old saying about "the best laid plans…." Plan A will always have a catch. It may be that a computer glitch that shows up just before you get ready to plot the final project plans, or the school baseball player who you're counting on to pitch calling in sick that day, or your star player not getting the grade needed to make the eligibility list for the game."

This chapter is about service to others, whether in the military, in business, or in volunteering. It's about being a leader in all facets of your life. It is about leading a life of selfless service and servant leadership. We believe that all of us from Bravo Company strove to live extraordinary lives, not only in our military careers, but our desire to continue to serve into our late years that include to these very days, 50 years later.

There is a story about Alexander the Great. It is said he was leading his great Army through a vast desert. His Army was strung out for miles with men and horses suffering terribly from thirst. Suddenly, a detachment of scouts came galloping back to the king. They had found a small spring and had managed to fill a helmet with water. They rushed to Alexander and presented him with the helmet. The Army held in place, watching. Every man's eye was fixed upon their commander. As the story is told, Alexander thanked his scouts for bringing him this gift, then without touching a drop, he lifted the helmet and poured the precious contents into the sand. At once, a great roar ascended, rolling like thunder from one end of the column to the other. The tale ends when a man was heard to say, "With a king like this to lead us, no force on earth can stand against us."

From the Ranger Creed: Acknowledging the fact that a Ranger is a more elite Soldier who arrives at the cutting edge of battle by land, sea, or air, I accept the fact that as a Ranger my country expects me to move further, faster, and fight harder than any other Soldier.

We would add this to the creed, "both during and after active military duty." Like all the Rangers highlighted in this book, we all carried forward the Vision of General Creigton Abrams' Charter—To change lives, in changing times.

15. HEROES AT HOME

Taking care of families is as important as taking care of soldiers. You are reenlisting a family when you reenlist a soldier.
— CPT Lawson Magruder

From left: Gloria Magruder, Jennifer Leszczynski, Kathi Picanco, Patti McNeme

WHEN LAWSON MAGRUDER COMMANDED A COMPANY OF PARATROOPERS in the 82nd Airborne in 1973 and a company of Rangers in 1975, no more than a dozen soldiers or 7% were married. It is far different today where 50% are married and close to 40% have children. Taking care of families has never been more important, not only because it is the caring thing to do, but because it can impact readiness if soldiers are distracted because of family issues.

Even though the 2nd Ranger Battalion had a small number of married soldiers, caring for its families was of the upmost importance to the battalion commander LTC Bo Baker and his spouse Betty. Bo directed when he formed the battalion that it would have family support groups at company and battalion levels. These groups would belong to the commander but would be run by dedicated spouses. Their primary purpose was to keep spouses informed and to provide support in time of need, particularly during deployments.

In Bravo Company's case, the company commander's spouse, Gloria Magruder, volunteered to lead the group which included officer and enlisted spouses. This group proved to be incredibly successful because of the dedication of Gloria and First Sergeant Block's spouse, Linda. They were a dynamic team who cared immensely for the spouses, their families, and single soldiers. They created a very inclusive environment where everyone felt valued. There was no rank consciousness among the spouses. They were all treated equally and on a first name basis. The Bravo spouses became a quasi-family unit that was founded on friendship and concern for one another. Strong personal relationships were built that have lasted a lifetime.

The extraordinarily successful Family Support Group model established in Bravo company would be "exported" to other units in the future by Gloria and two other spouses whose husbands served for decades in the Army: Jennifer Leszczynski and Kathi Picanco.

A fourth spouse, Patti McNeme, whose husband got out of the Army after his enlistment ended, stated this about her experience: "Since I was new to the Army life and new to Fort Lewis, I was not sure how to fit in. Gloria made that easy. So, in my life after the Army, I have tried always to greet people with a smile, be approachable, and make other ladies feel welcome, especially if they were new to an organization or neighborhood."

For each of these incredible ladies, their positive experience in Bravo company laid the foundation for their success as volunteer leaders in the future. Here is what these "Heroes at Home" whose biographies are found in the appendix, had to say about the impact of their time together many years ago:

From **Gloria Magruder** about her time leading the Bravo spouse group: "The environment was set from the top, thanks to Betty Baker and her husband Bo. Every spouse was important to Betty, and she treated us all with great love and respect, all the while making things fun in spite of the long working hours of our husbands. Rank was not an issue when it came to the needs of spouses in the unit. In Bravo Company, we worked hard to make all our spouses feel they were of value and what they were experiencing was shared. I think we were incredible cheerleaders for the soldiers, most of whom were very far from friends and family. So, we were their family."

From **Jennifer Leszczynski** whose husband Bill was a lieutenant: "As a new spouse, I was taught the value of shared absence and the spirit of helping one another. Spouses were taught not to expect their husband home at a prescribed time. We learned to do everything by ourselves or to help and offer advice to our sister spouses. We supported each other like Ranger buddies, and we were tight. I had several spouse role models in the battalion: women like Gloria Magruder, Kathy Abizaid, Patti Powell, Carol Malvesti, and Kathi Dubik. They were all extremely strong and very motivated women. They shared their knowledge and expertise with the younger wives."

From **Kathi Picanco** whose husband Joe was a young noncommissioned officer: "In Bravo company, the environment for spouses and families was great! Time was spent keeping the families informed as much as possible. We may not have known where they went and when they would come back, but the company was always available if we needed assistance. The great family support we received helped to keep the families going until our soldiers returned. I knew we would gather and keep each other company and support each other.

"Family support meetings were critical for all of us. It was time shared with other wives in our situation. It was a time to cry, and to help each other with the kids as well. I will never forget the Rangerette breakfasts at the dining facility. Those times were great beginnings of friendships and a realization that leaders did care about family!

"Without my friends from those days in the battalion, I am not sure that I would have made it through 31 years of Joe being in the Army. I applied what I learned into my roles within the different units Joe was serving and helped other spouses deal with military life. Gloria Magruder told spouses what was important and needed. She made us feel important to the Ranger family and we could count on help when needed."

From **Patti McNeme** whose husband took a five-year temporary leave of absence from an engineering firm to serve his Nation: "When the men were called away, many times in the middle of the night, we never knew where they were headed or when they would return. But life goes on and I had a job to do in taking care of our home and our three kids. Jim's frequent absences required us all to be resilient and adapt to whatever the day brought us. This experience and the confidence it brought me served me well after leaving the military in the civilian world in dealing with the problems of life. My husband's civilian job required him to travel overseas sometimes for extended periods of time, but I had been well equipped by our experience in the Ranger Battalion to handle life."

Gloria, Jennifer, and Kathi's spouses went on to serve at the highest levels during their careers. Over their decades of service, the percentage of married soldiers rapidly increased. The importance of care for families was magnified particularly during the past 20 years when the military's operational tempo reached its highest level ever. The importance of what are now called Family Readiness Groups has been magnified.

Responses from these four remarkable women reinforced several themes found in the previous chapters: service to something greater than self; courage; open, honest communications; adaptability and resilience; leading; lasting relationships; expanded worldview; and purpose in life. Here are their testimonials magnifying how these qualities developed early in their lives as a military spouse were carried with them the rest of their lives.

Patti McNeme emphasized service, courage, resilience, lasting relationships, and purpose in life: "The ladies in the Battalion had a support system that was invaluable. Because of the support of these ladies, it taught me the value of reaching out and helping, supporting others. Over the years after the military, I have been intentional in not only helping other ladies but also in participating in leadership of various women's groups. Most of these opportunities (but not all) came through our church where the Lord kept bringing ladies in distress to me. It has been a good life. In life, we all face trials and challenges, and some can be overwhelming.

"When our oldest daughter, Shannon died in a car accident at age 19, it was a very difficult time. It took Moral Courage and the hand of God for me to carry on. Once again, the Ranger support network jumped in to help us through our mourning. Even though we had been out of the military for many years, we received notes of sympathy from our Ranger acquaintances, including from Jim's first company commander, Lawson Magruder and his wife, Gloria. Shannon's death and the wonderful support I received, emphasized the need to 'give back.'

"I know that God has a purpose for all of us and I became even more intentional at helping others. I loved teaching young children in church and in the neighborhood. I also became more involved in leading women at our church. We all need a "Sense of Purpose and a Direction in Life" and I found mine."

Kathi Picanco emphasized open, honest communications, lasting relationships, trust, and teamwork, and leading: "One of the best theme/challenges would be the Lasting Relationships, shared responsibilities, and hardships. I believe in working in a group to get the tasks done and to listen to others' ideas on how to get tasks to completion. It is always helpful to hear how others would get the tasks done and why. None of us at an early age had all the answers or knew the effects that our decisions would have on our family members.

"All military spouses have had some kind of hardships to deal with during deployments, training missions, or other absences of their spouses. We all tried to help where needed—or were able. Shared Values & Building Trust & Teamwork is part of being a Support Leader. We all tried to draw on family members to participate in the support group. It was important for all family members to be able to voice what is important to them and their families. There were young spouses who needed more support than others, for them it could have been the first time their spouses were away. Having to set up a household or being a parent and managing it all alone was hard. Especially since it could have been the first time around the military.

"My biggest challenge later in life was to become the President of the Rebekah Assembly. As a statewide and International Officer for Rebekah's I had to talk to a large group of people, I am not always at ease speaking in front of a group of people, but again those early times in the Ranger Battalion and subsequent commands helped prepare me. I try to remember what I did as a Family Spouse Leader and a CSM Spouse and do my absolute best. It has been nice to have other people feel I can do the job as their representative. My husband has helped me to tackle this fear and just do it anyway. I have always volunteered even as a little girl. If I can, I want to help."

Jennifer Lesczczynski reinforced the themes of trust and teamwork, physical courage, leading, and service: "At this time in the Army, there was no Family Readiness Group or Family Support Group. However, Ranger wives just did what needed to be done. It was the beginning of the wives' buddy system. I learned many skills from the group of ladies mentioned above, but I learned to thrive when we had to function as a tight group, and that we did. We had the usual monthly coffees, exchanged recipes, company parties, and helped new moms with advice and babysitting.

If we had a serious question or problem we asked the more experienced ladies for advice.

"Our communication was always very good and that is one of the most important lessons I took with me throughout our career. If women knew what was happening, all would be much smoother. Throughout our time in the military, I enjoyed volunteering in many organizations i.e., OWC, PTA, church, and the unit family group but I still wanted to do something for me. I began running and eventually started running marathons. The self-discipline needed to finish a marathon was extreme, but it taught me to push through pain and discomfort to reach the goal.

"I would remember back when Bill was road marching in the rain at Fort Lewis, the extreme heat at Fort Benning with blisters covering his feet, thinking if he can do that, then I can finish this race, and I was always very proud to run in a Ranger Shirt! Over the years I recruited several wives to running and some of those women are still serving beside their husbands on active duty.

"In December 1989, Bill took command of the 3d Battalion, 9th Infantry (Manchus). Within days he deployed with his battalion to Panama for Operation Just Cause, the US invasion. He did not even know the names of all his staff and company officers, and I did not know the other wives. I was tasked to run our Family Readiness Group (FRG) and knew no one. I immediately set up a communication system and through that I met, and came to admire, many very smart and resourceful women. Even after redeployment, our family readiness group remained very tight. We had built something good, and we had suffered together.

"I have always been very proud of Bill's service to the Army, but most especially of his dedication to Rangers. I think we conveyed that to our children as well. Our daughter, Kristan, met and married a young lieutenant, Todd Brown, after meeting him in Vicenza, Italy. They've had multiple tours in the Ranger Regiment, beginning as a lieutenant in 2nd Ranger Battalion. Todd has served as the Executive Officer for 1/75, the Regimental Special Troops Battalion, and the 75th Ranger

Regiment. He also commanded the 1st Ranger Battalion and the 75th Ranger Regiment.

"They are currently stationed in Korea with their two sons. Todd is the Deputy Commanding General—Maneuver, for the 2nd Infantry Division. Kristan participated in the FRG in all those assignments, set up a babysitting co-op at Fort Benning, Georgia, organized a very successful auction fund raiser for their FRG, prepared countless meals for other spouses in need and raised two wonderful sons who are excelling in academics and sports. She performed these activities all while knowing her husband was in harm's way. The spouses of this generation have truly carried on the legacy of the Army spouses from the past. They have endured so much during the Global War on Terror and said final good buys to many friends who died in combat. They suffered in silence the entire 20+ years.

"Our son Pete also chose the military and has served multiple tours in the 75th Ranger Regiment, to include service in all three Ranger Battalions. He has served in Iraq, or Afghanistan, with each of the Battalions. He, his wife Katy, and their daughter are stationed in Hawaii where Pete commands 2nd Battalion, 27th Infantry (Wolfhounds). They will soon move to Joint Base Lewis-McChord and Pete assumed command of 2nd Ranger Battalion in June 2024."

"I am inextricably tied to 2nd Ranger Battalion. I find myself as a wife, mother, mother-in-law and potentially a grandmother of Rangers. Even though my husband is retired from active duty, and apart from the basic needs of life, i.e., eating, sleeping, and shelter, not a day passes when I don't think about our time in the Rangers and how tremendously lucky we were to have known such great American men and women. I am proud that we perpetuated the life of military service to the Nation and to our children, and that our children, and our grandchildren, will keep the story and tradition shaping the future."

Gloria Magruder emphasized the themes of courage, open honest communications, resilience, expanded world view, and purpose in life: "Going forward from the 2nd Ranger Battalion, this left me with a strong feeling of equality of all spouses in any military unit my husband was assigned to and especially when he was in a command position. As a parent and spouse, I became independent while remaining a complement to my husband as a spouse as he advanced in his career. I made it my mission throughout his career to never wear his rank and to always be referred to by my first name — never wanting to be known as the MRS.

"There are so many themes that I believe I either applied in life or I aspired to throughout my husband's career. One of those is "Adaptability & Resilience." I think being a wife in the Ranger Battalion and being a waiting wife while Lawson was in Vietnam, really helped me gain those qualities. I came from a broken family, with little to no understanding of the military, so when I married my husband, I truly did not know what I was getting into in this new lifestyle. But I just seemed to thrive and love the military, the people, the assignments, the challenges, and the joys. I had so many positive role models as senior spouses and then later in life, even a few negative ones that confirmed to me how I did not want to be. Those same qualities I believe we passed on to our three children as they moved numerous times, went to several different schools, and said 'goodbye' to many friends. As an example, our son went to four different high schools. Though none of them went on to serve in the military, they have a strong sense of patriotism and definitely are resilient and adaptable.

"During Lawson's career, I was profoundly blessed to experience some incredible opportunities to meet wonderful people from different cultures and countries. This definitely 'Expanded My World View and Respect for Others.' Of note, his assignment in Panama allowed me to meet some of the most amazing civilian employees who adored the soldiers and families of the military. Then I was also able to travel to other Central and South American countries with Lawson and met more amazing, strong, and courageous women from countries like Argentina, Colombia, and El Salvador. I loved those experiences and felt humbled and honored to represent our country and our programs as we visited these places. The people were so welcoming, loving, and interested in what we were doing to support our soldiers and their families.

"The military and the friends I made supported me as a wife and mother. It helped me with a 'Sense of Purpose & Direction in Life.' My faith was paramount in all posts we were assigned to and was a strong guiding force as I helped my children adjust to their new environment when we moved to a new post. This provided us all with the opportunity for 'Lasting Relationships & Shared Responsibilities and Hardships.' It was hard to say 'goodbye' to old friends but was quickly replaced by a welcoming new community.

"As Lawson rose in rank, I found myself in more positions of 'Leading Others.' I loved my military community and wanted the very best for our soldiers and their families, especially when soldiers were deployed. I made it my mission to work closely with the Garrison staff and the Director of Community Activities and worked hard to have those relationships be very positive with open communication and if a problem arose, with an understanding that I was coming to them as a concerned spouse, not wearing my husband's rank. I forged lasting friendships as a result of working closely with these individuals.

"As in all our lives, there are many challenges, but if I had to explain one, it would be a personal one that came later in life. I stated earlier that I came from a broken family, more appropriately called, a dysfunctional family system. There was extreme alcoholism, abuse, and lots of subsequent verbal altercations. This was my life from a very early age, to include abuse. But I managed to suppress it for most of my adult life. I had "stuffed" it deep in my psyche until such a point that I became severely depressed and had to seek help. I did this while balancing a family, managing to stay involved with my community, and supporting my husband.

"I had learned coping skills over the years that allowed me to get on with daily living and keep the past at bay, but even that took a toll. It wasn't until after my husband retired that I was able to focus my energy on getting the consistent, serious help I needed and getting well, once and for all. This is where 'Open & Honest Communication' was invaluable with Lawson. He was incredibly supportive of me as I untangled my abusive childhood, and he stuck by me in the toughest of times. He was rising to the occasion in no uncertain terms to be by my side while I dealt with my depression and my history. Without him and my strong background of resilience, my road to healing would have been a much different path.

"There is no doubt that my life in and around the military family had provided me with the stamina, the fortitude, and the courage to face the disease and the abuse and get well. I will forever be grateful to my husband, my family, and our military friends who supported me through this dark tunnel and my recovery to good health.

"In closing, I adored the soldiers and their families. I loved our time in the Army and felt very honored and humbled to be in positions where I could make a difference in the lives of those in my husband's command. It was truly a privilege and left me with memories and friends I hold dear to this day. The Ranger Battalion experience was truly laying the foundation for many of those positive experiences for which I am extremely grateful. 'Rangers Lead the Way' in so many areas of life, and so do their spouses."

* * *

Each of these special ladies has defined for us what being a good citizen means. Patti took from her time as an Army spouse a desire to help others in her Faith community and the families of Jim's large engineering firm. Gloria, Kathi, and Jennifer focused on raising incredible children and bettering their communities while helping their soldier spouse defend our Nation in peacetime and war. Kathi and Jennifer went the extra mile in their service to our Nation. Kathi not only served in uniform herself but had a son who was a proud military policeman who was wounded in combat and after retirement continues to serve as a DOD policeman. As described by Jennifer, she not only supported her spouse Bill in his military and civilian senior leader positions but has a daughter who is a proud

Army spouse and a son who is the current commander of the 2nd Ranger Battalion. Jennifer and Kathi have been "master recruiters" who set the example for so many others by encouraging their own to continue the family tradition of service.

May God bless the Heroes at Home worldwide who continue to make sacrifices in support of their soldiers who proudly serve our Nation.

EPILOGUE

THANK YOU FOR READING THIS BOOK. WE HOPE THE STORIES ABOUT OUR fellow Rangers have inspired you in some way to serve your country or community in some capacity. Perhaps you are a young American who is considering joining the greatest military in the world. Or a parent, coach, teacher, pastor, or influencer of our youth who want to encourage them to serve something greater than themselves. Or a serving military or civilian leader who gained some valuable leadership advice. Or a recruiter who found the themes helpful as you speak to potential recruits about the intangible benefits of military service. Or a veteran who reflected on your past experiences in uniform and now understands better the value of continued service and perhaps capturing and sharing your own story with others. Regardless of your position in life, we hope you will share this book with others.

What is most evident in these stories, is that individuals make a difference, especially in uncertain times. We hope, in these challenging and sometimes troubling days, these stories have served as a positive inspiration to get involved in something greater than yourself. We hope this book delivered a strong dose of encouragement to all.

To the young men and women who are thinking about joining the military to serve your country, here are six things to think about that might help you decide to make the honorable decision to join one of the branches of our great military.

Military Service will:

1) Make you better than you were before you served. More mature, more disciplined, more confident.

2) Build character, mental and physical resilience, drive and determination, positive attitude, strong self-esteem, personal pride, and a sense of honor and love of country—all the key elements to succeed in life.

3) Teach you selfless service by serving something greater than yourself while giving you new knowledge, skills, and abilities for a more successful post military life.

4) Build deep bonds of lifelong friendships.

5) Earn you tangible educational and medical benefits.

6) Allow you to become a part of a very small, elite club of just 1% of American citizens—the American Veteran.

For the men of Bravo Company, it was simply an honor to serve. We all wish we could do it again. Now it is another generation's opportunity to serve in changing times when your nation needs you.

In closing, we want to thank our spouses Gloria and Erika for their love, encouragement, patience, and initial proofreading. Also, fellow Vietnam infantry combat veteran, Bob Babcock and his Deeds Publishing team for their sage counsel and professional editing. Finally, we are grateful to our fellow Bravo Rangers for participating in this project. We value your transparency during the interviews and enduring friendship over the past five decades. It was our honor to capture your incredible stories. Rangers Lead the Way!

Lawson Magruder

Fred Kleibacker

BIOS IN ALPHABETICAL ORDER

First Sergeant (Ret) William D. Block: Bill joined the U.S. Army in December 1955. After completing basic and combat engineer training, he was assigned to the Headquarters Detachment, US Army Garrison in Paris. He changed his MOS to Infantry and successfully graduated Ranger School in September 1960. He served as an instructor for three years in the Ranger Department at Ft. Benning, GA. He volunteered for the 101st Airborne Division as a Reconnaissance and Rifle Platoon Sergeant. Bill did multiple combat tours in the Vietnam with the 101st AD and as an advisor to the Vietnamese 7th Airborne Battalion. Upon returning, he served another seven years in the Ranger Department. He volunteered for the 2nd Ranger Battalion in 1974 and was Bravo Company's original First Sergeant from 1975 to 1977 when he retired. He lives in Columbus, GA next to Ft. Moore (formerly Ft. Benning) and remains very active throughout numerous Ranger Associations, Meals on Wheels, and regularly visits hospitalized veterans and the Masons.

Captain (Ret) John Brasher: John joined the Army in 1975 spending 6.5 years as an enlisted member, reaching the rank of SSG. He went to OCS in 1981. He went to serve 20 years as an Infantryman before retiring as a captain in 1995. John's military schools include Airborne, Ranger, Jumpmaster, Pathfinder, OCS, Infantry Officers Basic Course, Infantry Officer Advanced Course, and Combined Arms Service Support School. His infantry assignments included, B Co, 2nd BN Ranger, 75th IN, C Co, 1st BN 509th ABCT, Ranger Instructor Camp Darby, Infantry Platoon Leader C Co, 1st BN 505th, 82nd Airborne, Company XO, CSC 1st BN 505th, 82nd Airborne, and Anti-Tank Company Commander E Co, 1st

BN 18th IN. After retiring he worked as a military contractor for 23 years in the Columbus, GA, Fort Moore area.

Colonel (Ret) Ron Buffkin: Ron earned his Ranger Tab as a nineteen-year-old PFC. Meritoriously promoted to Specialist the day the Army pinned the coveted black and gold tab on his shoulder, Ron served in every Ranger Rifle Squad position culminating as Squad Leader, 2nd Squad, 3rd Platoon in B Company. During his time in B Company, he also served on the hand-to-hand combat demonstration team, attended sniper training, Pathfinder school and was one of the first High Altitude Low Opening (HALO) parachutists in the battalion. Ron was commissioned in the Infantry through Office Candidate School (OCS) and volunteered for flight training earning Army Aviator wings in 1980. His career included command of attack helicopter units from company to brigade, commanding the Field Team for the Joint Improvised Explosive Device Defeat Organization (JIEDDO) in Operation Iraqi Freedom and he served in and commanded a Special Mission Unit (SMU). Ron retired from the Army as a Colonel in 2007 and worked as a defense contractor enabling aviation Foreign Military Sales (FMS). Ron has been married for 45 years with one son, a Marine veteran. Ranger Buffkin ascribes his 33 Army years to those superb peers and leaders from those first three years in B Company (75-78).

Sergeant Danny Crow: Danny joined the U.S. Army in April 1975. After he attended Basic and AIT he arrived at Ft. Lewis, WA in October 1975 without orders for a unit. He was a non-Airborne qualified private. While awaiting orders to the 9th INF DIV, he was interviewed and accepted into the 2nd Ranger Battalion. He was assigned to Bravo Company, 3rd Platoon. He attended Airborne School along with about 50 other leg Rangers from the Battalion shortly after arriving. He went on to attend Ranger School in October 1977. During his time in the company, he served as a Fire Team Leader and Squad Leader. He was also a platoon sniper and sniper team leader. He exited active service in 1978 and joined the California National Guard serving in the 76th Pathfinder Detachment in Stockton, CA. He attended the Police Academy in Modesto, CA and started working for Merced Police Department. He worked briefly for the Gustine Police Department before going to Tuolumne County Sheriff's Department. During that time, he was a Training Officer, SWAT team member, Search & Rescue team member and Firearms Instructor. Danny retired after 30 years of service.

Chief Warrant Officer 4 (Ret) Shaun Driscoll: Shaun joined the Army in 1975 under the Ranger Enlisted Option. He arrived at B Company, 2nd Ranger Battalion in October 1975 after finishing Basic, AIT and Airborne school. He served three years in B company as a Machine Gunner and Squad Leader in 3rd Platoon. He graduated from Ranger School in December 1976. Shaun reenlisted for Special Forces training in October 1978 and for the next 20 years served in various Special Forces assignments with 10th Special Forces Group as both a senior NCO and Warrant Officer. He deployed in Operations Desert Shield, Desert Storm, Provide Comfort (Iraq), and Silver Anvil (a non-combatant evacuation operation in Sierra Leonne, West Africa). As a senior Warrant Officer, he managed the SF Warrant Officer Program and was selected in 1995 to run the SF Warrant Officers Basic and Advanced Courses. He retired in 1998 and taught Junior Reserve Officer Training Corps (JROTC) in Newport and Barre, Vermont for 19 years. He has four grown children who all served in the Army. His oldest is still serving on active duty in the U.S. Army. Today, he is fully retired, living with my wife in Ft Walton Beach, Florida.

Lieutenant General (Ret) Jim Dubik: Jim was commissioned into the infantry in 1971 and retired from active duty in 2008. He was the second company commander of B Company, 2nd Ranger Battalion from 1976-1977. He also served in the 82nd Airborne, taught at West Point, became the XO of the First Ranger Battalion, commanded the 5th Battalion, 14th

Infantry, 25th Infantry Division and the Second Brigade, 10th Mountain Division during Operation Uphold Democracy. He served with the First Cavalry Division as Deputy Commanding General of Multinational Division North, Bosnia-Herzegovina. He created the Army's first Stryker Brigade Combat Teams. He commanded the 25th Infantry Division and I Corps. His last job on active duty was Commanding General, Multinational Security Transition Command-Iraq and the NATO Training Mission-Iraq during the Surge of 2007-2008. Jim earned a PhD in Philosophy from Johns Hopkins University and taught at Georgetown's Security Studies Program. He is a Senior Fellow at the Institute for the Study of War, teaches in the Hertog War Studies Program, and is a member of the Council on Foreign Relations, the National Security Advisory Council, and the U.S. Global Leadership Coalition. He is a U.S. Army Ranger Hall of Fame recipient. He is also a Board member and former Chairman of Leadership Roundtable, a Catholic non-profit assisting the Catholic Church in the U.S. in leadership and management practices. He was also the 2012-2013 General Omar N. Bradley Chair in Strategic Leadership at Carlisle, Pennsylvania. Jim authored Just War Reconsidered: Strategy, Ethics, and Theory and has published over 200 essays, monographs opeds, and articles in a variety of publications.

Sergeant First Class John Funderburk: John Funderburk joined the U.S. Army in December 1974. John was assigned to B Company, 2nd Ranger Battalion in April 1975 after completing Basic, AIT, and Airborne School. He served two years in B Company and then reenlisted for Special Forces. He served a combined three years in the 5th, 7th and 10th Special Forces Groups, before joining the 1st Special Forces Operational Detachment—Delta at Ft. Bragg, NC where he served for five years. He left the Army after 11 years and became a defense contractor for Bechtel in Algeria and Skylink Aviation in Afghanistan. Today John lives quietly in North Carolina with his wife and remains involved in helping folks out in his local community.

Sergeant Tom "Doc" Giblin: Tom Giblin arrived at Bravo Company, 2nd Ranger Battalion in December 1975 as an Airborne qualified Private First Class Medic. He left the unit in June 1978 as a Sergeant/E-5, Airborne Ranger, Jumpmaster qualified Medic. He returned home to begin his civilian career as an electrician in the Binghamton, NY area. He became a Master Electrician eventually starting his own business. He taught at the State University of New York at Delhi for 2 years. He worked as the electrical inspector for the City of Binghamton and finished his career at the Binghamton Johnson City Joint Sewage Treatment Plant. He worked there as both an electrician and as an operator running the plant. He is currently retired and spends his days helping folks hike the trails and valleys in the Catskills Mountains. He credits the training that he received as an Airborne Ranger in giving him a desire to never quit, never give up. He demonstrated that when in his 60's he had both hips replaced and then proceeded to summit the highest 35 peaks in the Catskills, during summer and winter.

Master Sergeant (Ret) Tom Gould: Tom Gould joined the Army in September 1973 for the Ranger Enlisted Option. He completed Basic, AIT, Airborne School and Ranger School and was assigned to B Company 75th Rangers a year before it was deactivated. Tom volunteered for the newly activated 2nd Ranger Battalion. He was assigned to B Company, 1st Squad, 3rd Platoon as a team leader. He reenlisted for Special Forces in 1978. He served in the 10th Special Forces Group and volunteered in 1980 for Detachment A in West Berlin, FRG during the cold war. He was reassigned in 1984 as an instructor with the Special Forces Qualification Course at the Special Warfare Center training at Ft. Bragg, NC. He was selected in 1986 for the 1st Special Forces Operational Detachment—Delta at Ft. Bragg and remained there until his retirement in 1994. Tom started the first Junior Reserve Officer Training Corps (JROTC) program in Montana. Tom retired from JROTC in 2006 and went to work as a Private

Military Contractor. He worked in Africa until 2016. He is fully retired today with his family in Montana.

Sergeant Tim Grzelka: Tim Grzelka joined the Army in October of 1972 on the Ranger Enlistment Option in Beaver Falls, PA. He completed Basic Training, Advanced Infantry Training, Airborne School and Ranger School in the summer of 1973. He was assigned to Company B Ranger, 75th Infantry at Fort Lewis, WA. In 1974 he volunteered for the 2nd Ranger Battalion and was assigned to Bravo Company, 1st Platoon,1st Squad as the Squad Leader. He departed the Army in February of 1976 and took a job in the medical field repairing hospital equipment. He graduated from Penn State University with a degree in Biomedical Engineering and spent the next 29 years as a medical equipment service technician for hospitals and home health facilities in the Pittsburgh, PA area. When he retired in 2005, he and his wife relocated to Bozeman, Montana to enjoy the mountains, fishing and hunting.

Captain (Ret) Dave Hill: Dave Hill enlisted in the Army in October 1974 for the Ranger Enlisted Option. He arrived at Bravo Company, 2nd Ranger Battalion in Ft. Lewis, WA in March 1975. He served for two years as a mortarman before leaving active duty to attend college in October 1977. He joined the Maryland National Guard serving in the 11th Special Forces Group while attending college. He graduated in May 1982 and was concurrently commissioned as an infantry officer. He served in a variety of active duty operational and training assignments, both overseas and in the U.S. He retired in 1994 and accepted a civilian position at the Joint Readiness Training Center, Operations Group in Fort Polk, LA as Chief, Training Analysis Computer Support and Simulations. In 2005, he accepted a position as the Joint National Training Capability (JNTC) Certification Manager with US Joint Forces Command (USJFCOM) in Suffolk, VA and subsequently under Joint Staff, J7. Dave continues to serve in the Joint Staff, J7 with the Service Joint Training Division in Suffolk, VA. Dave lives with his wife Chrystie in Newport News, VA. They have been married for 34 years; no kids but live happily with an energetic Jack Russell terrier named Luke.

Major General (Ret) Jim Jackson: Jim served in the Army and the special operations community for 32 years. He commenced his career in the 82nd Airborne (ABN) Division (DIV) and arrived at Bravo Company, 2nd Ranger Bn in 1975. He served as a PL and XO for B Company, Battalion S3 Air, and Commander (Cdr) of Charlie Company. Following assignment to Korea as Bn S3, 1-23 Infantry, he returned to the 82nd Abn Div as the 3d Brigade (Bde) S4, S3, Bde XO and Cdr of 1st Bn (Abn) 505th INF. He became the G3 Ops Officer for the 7th INF Div during Operation JUST CAUSE in Panama. He commanded the 3rd Ranger Battalion. He served as USSOCOM Plans officer before accepting command as the Eighth Colonel of the Ranger Regiment. During his command the Regiment participated in combat operations in Somalia and Haiti. After command he served as the Deputy J3 at USSOCOM and later the XO of SOCOM. He returned to Korea as the Assistant Division Commander for the 2nd INF Div followed by Deputy Fifth Army Commander. His final duty assignment as Commanding General of the Military District of Washington included security and support of operations following the terrorist attacks on 11 Sep 2001.

Master Sergeant (Ret) Alan Kovacik: Alan Kovacik joined the Army in February 1973 under the Airborne Ranger Enlisted Option. Completed basic, AIT, jump school and Ranger school and was assigned to B Co, Rangers, 75th Infantry at Ft Lewis, WA, Oct 73. He served with the 75th Rangers until it was disbanded and then was reassigned to newly activated Bravo Company, 2nd Ranger Battalion in 1975 until June 1978. Al re-enlisted in June 1978 to become a Drill Sergeant and later at the Ranger Department as an instructor. He departed active duty in December 1981 and took an assignment as a reserve drill sergeant at Ft Leonard Wood. He rejoined the service in Nov 1983 as an Intelligence Analyst and served in multiple units both overseas and the U.S. until his retirement in 1996. After retirement, he worked as a civilian contractor for General Dynamics as an instructor in Ft. Wainwright, AK. In 2022 he retired to hunt and help raise 8 grandchildren with a ninth on the way.

Brigadier General (Ret.) William J. Leszczynski, Jr.: Bill graduated from the United States Military Academy. His initial assignment was the 82nd Airborne Division before joining 2nd Ranger Battalion in 1975 where he served as a rifle platoon leader and company executive officer. He subsequently served as Executive Officer, 2nd Ranger Battalion, and Commander, 75th Ranger Regiment. Other assignments include Commander, 3d Battalion, 9th Infantry Regiment; Commander, JTF-Bravo, Honduras; DCG, USASOC; and ADC(S), 82nd Airborne Division. After retiring, Bill served as Director, European Region, and Executive Director, American Battle Monuments Commission. Bill and his wife, Jennifer, have been married for 50 years. They have two children. Their daughter, Kristan, is married to Colonel (P) Todd Brown, who commanded 1st Ranger Battalion and the 75th Ranger Regiment. Kristan and Todd have two sons, Joey and Jack. Their son, Pete, his wife Katy, and daughter Charlotte Grace, are currently stationed in Hawaii where he commands 2nd Battalion, 27th Infantry (Wolfhounds). Pete has served in either Iraq or Afghanistan with the 1st, 2nd, and 3d Ranger Battalions and assumed command of 2nd Ranger Battalion in June, 2024. Bill's quote: "My first time in the 2nd Ranger Battalion was the best and most important assignment I ever had."

Jennifer Caroll Leszczynski: Jen was born in Fort Bragg, North Carolina, the daughter of a career Artillery Officer, and Korean War Veteran. She holds a degree in History and Secondary Education. She also attended the Defense language institute in Monterey, California for Italian language and Alliance Francaise for French language training. She taught school upon graduation from university. She became an adult education counselor for the Army Education Center at Fort Bragg. After transferring to Fort

Lewis, she worked in the Office of the Comptroller, Fort Lewis, Washington. Jen participated in several humanitarian relief efforts during this time. After transferring to Italy, she taught English as a second language to Italian officers in the Carabinieri (Italian Federal Military Police). for two years. After returning to the USA, she volunteered her time to various organizations such as, PTA, Catholic Church, school, AFTB, OCS, FSG (FRG), Leadership Seminars, President, Fort Benning, GA and several humanitarian projects. After her husband left the army, and began working for the American Battle Monuments Commission, as Director of the European Region, they were posted in Paris France. Jen began working at the American Embassy, Paris, and was involved with the French Officers Wives group in Paris. Returning to the USA, and residing in Northern Virginia she ended her work career working as a government contractor with the Pentagon Force Protection Agency. She has been married for 50 years, moved more than 26 times has two children, Kristan Brown and LTC Peter Leszczynski and four grandchildren. She and her husband now reside in Mid Coastal Maine.

Command Sergeant Major (Ret) Clifford C. Lewis: Cliff joined the U.S. Army in July 1974. After Basic, AIT, Airborne and Ranger school Cliff arrived at 3rd Platoon, Bravo Company, 2nd Battalion, 75th Infantry (Ranger) in June of 1975. While assigned to Bravo Company his duty positions included: Rifleman, Team Leader, Squad Leader, Weapons Squad Leader, Anti-Tank Section Leader, and Weapons Platoon Sergeant. Cliff's education includes Bachelor of Science in Business Management. Cliff retired in 2000 with 26+ years of active service. Cliff and his wife are retired in the Ozarks in southern Missouri where he is active in his local New Testament Church as a Deacon and Treasurer. Cliff's quote about his service "The Leadership and Training in Bravo Company wasn't just special, it was exemplary. Truly Elite!"

Colonel (Ret) Mark T. Lisi: Mark enlisted in the Army in November

1974 for Ranger Enlisted Option unit of choice, the 2nd Ranger Battalion. He arrived at Bravo Company, 3rd Platoon in April 1975 after completing Basic, AIT, and Airborne School. Mark served for three years before deciding to return to college in 1978 and participate in ROTC. He graduated and was commissioned a 2nd Lieutenant of Infantry in May 1980. He returned to active duty and served until 1992 at which time he joined the reserves and went back to school to obtain a Public School Teaching Certificate. Mark was a public school teacher in Tacoma, WA for six years. He was mobilized in 2004 for Afghanistan and retired in 2006 after 24 years of service in both the active and active reserve components. He served in various leadership and staff positions as an Infantryman: Airborne Rifle Platoon Leader, Support Platoon Leader, Battalion Adjutant, Battalion Logistics Officer, Rifle Company Commander, and Battalion Operations Officer. Mark served overseas in Alaska, Republic of Korea, he first Gulf War and Operation Enduring Freedom in 2005. He is retired and living with his wife in Sparks, Nevada.

Gloria A. Magruder: Gloria has been married to Lawson W. Magruder III for 55 years now. She is the proud mother of two daughters and one son and the grandmother of four amazing grandchildren. Attended the University of Texas until she married in 1969 then went back to college in 1986 to change her degree from Business to Criminal Justice. Gloria has held many volunteer positions: Worked as a Rape Crisis Intervention specialist while in college; served on numerous Boards over her husband's 32-year career to include but not limited to the Red Cross, Thrift Shops, and Army Community Services; developed a victim-witness advocacy program with the SJA in the 25th Infantry Division, Hawaii; facilitated workshops with Army Family Action Plan and Army Family Teambuilding; helped develop the Company Commander's Spouses workshops at Ft. Benning (now Ft. Moore), Georgia; served the community at large installations during her husband's command tours; worked as the Director of the Armed Services YMCA in Hawaii for 3 years; went to Massage

Therapy school in Atlanta in 2000 and then worked as a Massage Therapist for several years after Lawson's retirement; volunteered as a CASA (Court Appointed Special Advocate) for abused and neglected children; and is presently volunteering at St Anthony de Padua Church in San Antonio, Texas and in the Army Residence Community where they live.

Staff Sergeant Kim Maxin: Kim Maxin was drafted in April 1971. He completed basic, AIT, and airborne school and was assigned to Company B, 75th Infantry (Ranger), at Ft. Carson, CO. The unit moved to Ft. Lewis, WA, and formed the core of the newly activated 2nd Ranger Battalion. Kim volunteered and was assigned to Bravo Company, 3rd Platoon, 3rd Squad Leader in 1974. In 1978 he took an assignment in the 25th ID. When his enlistment was over, he decided to leave the military and go to college, receiving his bachelor's and master's degrees. Kim spent the next 30 years as a school counselor helping young people with their career decisions and to serve his community as a school board member, president of a local chapter of school counselors, and as president of a sports booster organization which he helped establish. He retired in 2018 and currently volunteers to help seniors and stay in their homes. Kim said of his Army experience that it changed his life forever and that his time with the Rangers, especially the 2nd Battalion, helped him grow as a leader and to learn the deep value of teamwork, service to my country, and to gain lifetime friendships.

Command Sergeant Major (Ret) Ricky J. McMullen: Ricky enlisted in September 1974 from Alaska, where his father was currently stationed at Ft. Richardson. Upon graduation from Airborne School, he was assigned to the 2nd Battalion and arrived in January 1975. He served 2.5 years in B Co and re-enlisted for an ABN assignment at Ft. Richardson, AK. Rick went on to serve in Airborne Infantry and Mechanized and Light-Infantry units, as well as Instructor assignments throughout the Army Training Command. Rick finally retired in 1998, but as a Civilian continued in the training arena, working as an Instructor for the Department of Corrections, the State of Missouri and later as an Instructor/Manager for an Army Contractor and Civil Service. Rick and his wife now live on a hobby farm in Missouri spending time with his three Grandsons and is actively involved in his small country Catholic Church. He currently has a son carrying on the family tradition, with over 18 years of service as an Airborne CW3 Intelligence Analyst. Rick accredits his initial assignment in B Company, as "THE" role model in strong Character Leadership from the Team Leader to Company Commander position (Garrison & Field) and used those sterling and prominent examples as a template, a measure in "How it should be" throughout his military career.

Sergeant Jim McNeme: Jim joined the U.S. Army in 1974 after taking a leave of absence from his work where he was Manager of Engineering and Sales. After completing basic, AIT and Airborne School, Jim was assigned to B Company, 2nd Ranger Battalion in February 1975. He served 3 years in B Company, with 2 of those years in the position of Training NCO. As the Training NCO, he brought realistic and exciting training exercises to the unit. After leaving the Army, he returned to his former employer in Houston and served as Division Manager for an international Engineering and Construction company where he was responsible for about 800 employees. After the company was sold, Jim started his own engineering and construction company which he ran for over 30 years. He specialized in structural engineering and building difficult construction projects. After being self-employed for over 30 years, he retired at the age of 79. Jim attributes the success of his company to the values he learned in the Ranger Battalion. He only hired employees who were dedicated to excellence and teamwork. He trained them on how to excel in their jobs and they traveled the world and made key decisions with confidence.

Patti B. McNeme: Patti was born and raised in the Dallas, Texas area where she met her future husband, Jim, at the age of 13. They married at age 20 and recently celebrated their 60th anniversary. When they married, Jim was an engineering student at S.M.U. So, she put her educational plans on the shelf and worked to help put Jim through college. They were blessed to have three children and now have three "extremely bright" grandchildren. Patti soon realized she had a passion for helping others, primarily women and children. She has organized and served as Chairperson of the Ladies Ministry for three different churches, where the emphasis was on the building of relationships between the women. Patti went through special year-long training to become better qualified to help women. She learned to walk alongside a hurting person and provide one-to-one, emotional and spiritual care and connect that person with God's love. Because of her passion for children, she was blessed to have taught young preschool children for over 40 years. She also helped prepare and deliver lunches to children left at home while their parent(s) worked. To summarize, Patti loves being a wife, a mom and a helpmate to others.

Master Sergeant (Ret) Andrew "Andy" Pancho: Andy enlisted in January 1973. After Basic he was assigned to Ft. Lewis, WA to a conventional infantry unit for his infantry AIT. During his training he learned about B Company, 75th Infantry (Ranger) on post and applied. Andy was selected and attended Airborne School in June 1974. He volunteered for the 2nd Ranger Battalion when it was activated, and B Company Rangers was deactivated. He was assigned to Bravo Company, 2nd Ranger Battalion in 1975 and attended Ranger school in March 75. He served as a team leader in 1st squad, 2nd platoon. He left the Battalion in June 1976 and served in many leadership positions in Mechanized and Infantry units. Andy returned for a second tour with C Company, 2nd Ranger Battalion as a Squad leader and Platoon Sergeant. He had the privilege to serve as Drill Sergeant at Ft. Benning, GA, and as a Ranger Instructor training future soldiers and Rangers. Andy retired from the Army after 22 years, but his fondest memories were times spent in the Ranger Battalion alongside great leaders and soldiers.

Command Sergeant Major (Ret) Joseph Picanco: Joe entered the Army on March 1973 under the Ranger Enlisted Option. He completed basic, AIT, Airborne School and Ranger School and was assigned to Co B, 75th Infantry (Ranger) at Fort Lewis, WA in October 1974. He volunteered for the 2nd Ranger Battalion in 1975 when B Company was deactivated and was assigned to Bravo Company, 1st Platoon, 2nd Ranger Battalion. He served as a squad leader until 1977. Joe went on to serve for another 26 years in combat infantry units, including light, mechanized and Airborne Infantry, both overseas and in the U.S. He returned to his beloved 2nd Ranger Battalion as the First Sergeant (1SG) of Headquarters Company and Alpha Company. After stints with ROTC and the Joint Readiness Training Center (JRTC), he served overseas with several different Airborne Infantry units before retiring from the Cadet Command as the Command Sergeant Major in August 2003. Today Joe is a Grand Master in the "Independent Order of Odd Fellows," an international non-profit charitable organization. He is married to his life partner and wife Kathi of 48 years. Joe and Kathi have four grown sons and four grown grandchildren.

Kathleen R. Picanco: Kathleen (Kathi) Picanco was born at Madigan Army Hospital, Fort Lewis, Washington. Her father was a career-enlisted soldier. She spent three years on active duty with the U.S. Army and ten more years with the Army Reserve and Army National Guard. Her education includes a Master of Science in Personnel Management, 1982, from Troy State University at Fort Benning, Georgia. Kathi has retired from DOD Civil Service after 30 years in a variety of jobs. Her awards include Outstanding Volunteer Award 1996 and 1997, Commanders Award for Public Service, Four Official Commendation Awards, Employee of the Quarter and Employee of the Year from Fresno Military Processing Station. Her community service contributions include Family Support Groups where she has served as a Contact Person, Co-Coordinator and as an Advisor and Coordinator in various units.

She has served as an Advisor, Membership Chairperson, Treasurer, President and Vice President in various Enlisted Spouses Clubs, Non-Profit Organizations and Sergeant Major Wives Association. Worked as a volunteer in the Army Family Team Building, Family Assistance Center, ACS, Red Cross, Family Force Forums, Women's Conferences, School Enrichment Programs, Thrift Shop Council, Community Advisor Council, VA Hospital, and a Cub Scouts Den Leader. She has guest facilitated in three Pre-Command Courses at Fort Leavenworth, Texas.

Kathi for the past 48 years has been married to Command Sergeant Major Joe Picanco and is the mother of their two children, Jason 47, and Christopher 46.

Master Sergeant (Ret) Roy "Rip" Prine: Rip joined the U.S. Army June 1968. After Basic, AIT and Jump School he attended Special Forces Training and was assigned to the 6th Special Forces Group. He graduated from Ranger School in 1970 and then went to serve with the 5th Special Forces Group and later as an instructor at the Special Forces Underwater Operations. He volunteered for the 2nd Ranger Bn in 1975 and arrived at Bravo Company in January 1975. He was assigned to Weapons Platoon as Platoon Sergeant. In 1978 he departed the battalion and served with 7th and 10 Special Forces Groups. He completed his military career as an ROTC Instructor. After retirement he was a sergeant with the Florida Department of Corrections. He later graduated with a Degree in Journalism and worked as a reporter. He lives with his wife in rural Florida along with their two cats, Zorro and Hondo. He is a Lay Leader with his church and active member with the American Legion. "My time with the 2nd Ranger Bn was invaluable to me. It honed my leadership to higher degrees which I have used throughout my life's endeavors."

Sergeant Brian Quinlan: Brain volunteered for the "US Army Ranger Enlistment Option" on 7 July 1974, Brian completed Infantry basic, AIT, Jump School and Ranger school and assigned to Bravo Company, 3rd Platoon, 2nd Ranger Battalion. Brian spent two years in the Battalion as a senior rifleman, team leader, and a squad leader. He left the Battalion in 1977 and went to the 9th Infantry Division as a Squad Leader in an infantry platoon. Brian left the Army in 1978. He went home to the job he had before the Army as a welder. In the late 70's and early 80's, there was a lot of commercial nuclear power plant construction in various places in the country. Brian began to specialize in nuclear construction and quickly rose from a welder to a project manager for both new construction and operating maintenance of nuclear plants. He ultimately ended up specializing in inspections and project management. He has 44 years of experience in nuclear plant construction, maintenance and inspection. Today Brian is currently managing two new construction projects in Georgia. Looking back on his years in the Ranger Battalion he said, "The training, experience, knowledge gained, and personal friendships would be of benefit for the rest of my life."

Colonel (Ret) Marshall Reed: Marshall was commissioned as a Second Lieutenant in the Regular Army through ROTC at Stephen F. Austin State University in May 1973 and retired as a Colonel in January 2001. He served three years in the 2nd Ranger Battalion in both B Company and Headquarters Company. He also served in the 4th Infantry Division, 7th Special Forces Group, the 82nd Airborne Division, the US Army Infantry Training Center, 24th Infantry Division, US European Command and the US Army Combined Arms Center. After retirement he worked as a contractor in support of a series of Project Managers who were responsible for fielding command and control systems to Army headquarters from battalion through Theater Army. He and his wife live in Steilacoom, WA, about three miles as the crow flies from the original 2nd Battalion area on North Fort Lewis. The three years he spent as a young Ranger in 2-75 poured the foundation for his life as a professional soldier.

CSM (Ret) Darby Reid: Darby joined the U.S. Army in the summer of

1974, and was assigned to Weapons Platoon, B Company, 2nd Ranger Battalion in April of 1975. He served on a 90mm Recoilless Rifle crew and then as the Battalion Ammunition Sergeant. He departed active duty in October 1979 and joined the California National Guard in 1980. He joined the San Francisco Police Department (SFPD) in 1981. In 1984 he transferred to the U.S. Army Reserve. He retired from the SFPD in 2012 as a Sergeant after having served as a patrol officer, a plainclothes officer, a sniper, a field training officer, a SWAT officer and finally a patrol supervisor (sergeant). He retired from the U.S. Army Reserve in 2015. "The training and mentorship I received in the Ranger Battalion was what made me successful in the variety of assignments I had in the Police Dept, and in the Army Reserve."

Brigadier General (Ret) Jim Schwitters: Jim enlisted in the Army as a radio teletype operator (RTO) in 1975 for the 2nd Ranger Battalion. After initial training he was assigned to Bravo Company as the company commo chief and commander's RTO. In 1977 he was selected as a charter operational member with the 1st Special Forces Detachment—Delta where he served until his commissioning in the Infantry in 1981. Jim served in multiple positions in several infantry and special operations units, culminating as the Delta commander in 2002. As a General Officer he served in several assignments in Afghanistan and Iraq and completed his service as Commanding General of Ft Jackson, where he pioneered the implementation of Outcomes Based Training (OBT) in initial entry training. Since retirement, Schwitters has assisted other former colleagues with expanding the use of OBT in other military settings and law enforcement. Since 2013 he has been the master aboard a river tugboat in the riverine system and salmon fishing grounds of Bristol Bay, Alaska. Schwitters' quote about his time in B/2-75: "There could be no finer place than B Company for a newly enlisted Soldier to learn the power and capability of a military organization comprised of motivated disciplined Soldiers, skilled NCOs and proficient commissioned leadership. I learned there how to both follow and lead; lessons I carried through every subsequent assignment."

MSG (Ret) Jim Smith: Jim joined the Army in May 1975 at 19 years old. After initial training he was stationed at Ft. Lewis in hopes of joining the 2nd Ranger Battalion. He was assigned to Bravo Company, 3rd Platoon. He was sent to Airborne School and later he graduated Ranger School. Jim served as a rifleman, team leader and squad leader during his time in the Battalion. He was reassigned to Fort Benning, GA, in 1981 as a Ranger School Instructor and Operations Sergeant. He was reassigned to 25th Infantry Division at Schofield Barracks, Hawaii, where he served for 10+ years in various instructional, leadership, and staff positions. His final assignment was as Sergeant Major (Senior Instructor) at the University of Oregon ROTC Detachment. Since retiring in 1996, Jim owned and operated a small business, is very involved with Boy Scouts serving in state level executive positions and has been recognized for his achievements multiple times serving on his School District board as Vice Chair and Chairman for multiple terms. He also serves as a Trustee at his local Veterans of Foreign Wars (VFW) Post 4166. Jim and his wife Kristi have seven children.

CSM (Ret) John Snape: John joined the Army in 1967. After completing basic and AIT he was sent to Germany for one year. He volunteered for Vietnam and did two combat tours between 1968 and 1971. He returned to Ft. Ord, CA, and served as a Drill Sergeant program from 1971 to 1974. Reassigned to Ft. Lewis, WA. He joined the 2nd Ranger Bn and was assigned to Bravo Company serving as a section Leader for Weapons Platoon and Platoon Sergeant for 3rd Platoon. In 1977 he was posted to Ft. Richardson, AK, and served as an Airborne Infantry Platoon Leader. In 1980 he volunteered to be a Ranger Instructor and moved to Ft. Benning, GA. In 1984 John was promoted to First Sergeant and was reassigned to a Basic Training Company. He was then reassigned to help start up the Joint Readiness Training Center, located in Little Rock, AK. He was promoted CSM of the 3rd Battalion, 17th Infantry, 7th Light Infantry Division at Ft. Ord, CA. He retired after 27 years and worked for the State of California's Employment Development Department helping veterans find employment. He and his Wendy are fully retired and living very happily Oceanside, CA.

Chief Warrant Officer 2 Bill Waterhouse: Bill joined the army in 1973 at Fort Lewis WA, for B Company (Rangers), 75th Infantry. After it was deactivated in 1974, he reenlisted for the newly activated 2nd Ranger Battalion. He was assigned to Bravo Company, 2nd Platoon, 3rd squad, as the first Alpha Fire Team leader. He graduated from Ranger School in 1976 and was promoted to E5. He left the Battalion to serve three years with the 3rd Infantry Division in Germany. In 1980 he completed the Warrant Officer flight program and was assigned to the 5th ID at Fort Polk Louisiana. After completing the A1 Cobra helicopter transition course, he was assigned to the 4/12th Cavalry until 1986 when he left the Army to join the California Highway Patrol. He spent the next 14 years as a successful patrol officer and pilot until an injury forced his retirement in 2001. He spent the rest of his career as an Emergency Room/ ICU technician retiring in December of 2021. Still actively serving in his church, he now occupies much of his time with his sweet bride of 32 years, their 5 children, 13 grandchildren, and 2 great-grandchildren.

MSG (Ret) John Welgos: John joined the Army in 1971 for the 82nd Airborne Division. After completing Basic, AIT and Airborne School, he joined the Division at Ft. Bragg, NC. He attended Ranger School in 1972 and reenlisted for recruiting duty near his hometown in NJ in 1974. In 1975 John volunteered for the 2nd Ranger Battalion arriving at Ft. Lewis, WA in January 1975. After two years in the Battalion, John attended the Special Forces Qualification Course at Ft. Bragg, NC, and was assigned to the 1st Battalion, 10th Special Forces Group in Bad Tolz, Germany. In 1979 he left active duty, joined the reserves and became an Edison New

Jersey Police Department patrol officer. He rose to the rank of Lieutenant and eventually commanded the narcotics squad and the SWAT team. After responding to the 9-11 terrorist attacks in New York City, John retired from the department to serve his country again. John did multiple tours of duty as both a defense contractor in Gaza and Iraq and as a Special Forces augmentee in Afghanistan. He was medically retired in 2010 after 32 years of service. John lives with his wife Karen in NJ where he continues to serve veterans through a variety of veteran organizations.

Sergeant Mark Wheeler: Mark enlisted for the 2nd Ranger Battalion under the Army's Delayed Entry Program in December 1975, halfway through his Senior year of high school. After graduating Basic, AIT and Airborne School, Mark arrived at Ft. Lewis, WA, in November 1976 and was assigned to 1st Squad, 2nd Platoon, Bravo Company. During the following 39 months, he was promoted from PVT-2 (E-2) to SGT (E-5), attended Ranger School (Class 1-78), Jumpmaster, Military Freefall (HALO), Special Operations Training, Jungle Operations Warfare School, Primary NCO Course, Basic NCO Course, received his Master Jump Wings and Expert Infantryman Badge. After departing Active Duty, he spent 4 years in the Active Reserves with the 12th Special Forces Group at Sand Point Naval Air Stations, WA, and was promoted to Staff Sargent (E-6) as the Team Commo Sgt. Currently, Mark and his wife Paula of 44 years live in Bonney Lake, WA, just east of Ft. Lewis (JBLM) where they own a Mechanical Contracting/Design business. They have been blessed with a daughter & son-in-law, a son & daughter-in-law and seven grand-kids with one more 'standing in the door.' Mark's thought for the day, "You can only play RANGER for so long until you become one, then it's yours forever."

Colonel (Ret) Robert "Bob" Williams: Bob enlisted in the Army in 1975 for the 2nd Ranger Battalion. After Basic, AIT and Airborne School he was assigned to Weapons Squad, 3rd Platoon, B Company as a machine gunner. He completed Ranger School in 1976 and moved to a rifle squad until he left active duty in 1978. In 1979 he started college and joined the Michigan Army National Guard and was commissioned after graduating from the state Officer Candidate School. He went back on active duty as a Captain and went to Germany. He served in multiple units for over 13 years. Bob retired as a Colonel from Heidelberg in 2005 after 30 years of service. He now resides in the Upper Peninsula of Michigan where he enjoys the outdoor life, is an author, and works part-time supporting the US Army's War Fighter Exercise program. Bob reflects on his time in the Battalion, I consider them as my "informative years" ... setting the cornerstone for the rest of my time with the Army. One of the things I cherish most from my time in a Ranger Battalion, is the bond that extends beyond the generations. If you meet another Ranger Battalion soldier, past or present, that special bond is always there.

Captain Joe Wishcamper: Joe Wishcamper graduated from West Point in 1972 and was commissioned as a 2nd Lieutenant of Infantry. After graduation from Infantry Officer Basic Course (IOBC), airborne and Ranger schools, he was sent to his first permanent assignment at Ft. Carson, CO, where he served as a mechanized infantry platoon leader, a Rifle Company XO, a Support Platoon Leader, and a Battalion Support Officer (S-4). In 1975 he was selected for assignment to the then, newly formed 2nd Ranger Battalion at Ft. Lewis, WA. Joe served as the platoon leader of 1st Platoon, B Co. under the leadership of Captain Lawson Magruder, who was the first (plank holder) commander of Bravo Company. Joe was the plank holder of 1st Platoon Leader for approximately 2 years until he was promoted to Captain and was reassigned to a position in the 9th Infantry Division. He resigned his commission in 1977 and went to law school at the University of Washington in Seattle, graduating with his J.D. degree in 1980. Joe then practiced law in Bellevue, WA, for approximately 37 years until he retired in 2017. He and his wife Susan reside in Snohomish, WA, and are currently contemplating a permanent move to AZ at some point in the next year.

GLOSSARY

11B (pronounced 11 Bravo): U.S. Army Enlisted Infantry Military Occupational Specialty (MOS).

160th Special Operation Aviation Regiment (SOAR): Formed due to the failures of Operation Eagle Claw to rescue American hostages held by the Iranian government in 1980. The SOAR is considered a highly secretive Aviation Asset.

1st Special Forces Detachment—Delta (1st SFOD-D): Anecdotally known as the Delta Force. It is a storied Special Mission Unit formed in 1977 to combat terrorism and is the most elite special operations unit in the U.S. Army.

75th Ranger Regiment Association: An association for all Rangers who served in U.S. Army Ranger Units since the Korean War. They encourage fidelity, fellowship and camaraderie.

90mm Recoilless Rifle: Koren and Vietnam Wars era anti-tank and anti-personnel crew served weapon used in the post, Vietnam War era throughout the 70's and mid 80's, before being replaced by lighter, more powerful, more accurate anti-tank weapons.

Abrams Charter: In January 1974, General Creighton W. Abrams Jr. Chief of Staff of the Army, directed the formation of the first two Ranger battalions. The 1st Battalion (Ranger), 75th Infantry, was activated and parachuted into Fort Stewart, Georgia, on July 1; the 2nd Battalion (Ranger), 75th Infantry followed with activation on October 1, 1974. The modern Ranger battalions owe their existence to Abrams and his charter: "The battalion is to be an elite, light, and the most proficient infantry in the world. A battalion that can do things with its hands and weapons better than anyone. The battalion will contain no 'hoodlums or brigands' and if the battalion is formed from such persons, it will be disbanded. Wherever the battalion goes, it must be apparent that it is the best."

Active Guard and Reserve: The Active Guard Reserve (AGR) program allows Soldiers, both Officers and NCO, transitioning off active duty, the opportunity to compete for AGR positions closer to their home and remain on active duty with all pay and benefits. The program supplements active units who have manning requirements, often on deployments overseas.

Advanced Individual Training (AIT): AIT is applicable to all Military Occupational Specialty (MOS) in the Army. After Basic Training, all soldiers attend AIT for the specific occupation for which they enlisted. 11B was the primary MOS in the Rangers, which was a combat infantryman.

Aerial Rifle Platoon: These platoons were specific to the Vietnam War and to Air Calvary units like the 101st Airborne and 1st Calvary Divisions. They were infantry platoons that were 100% mobile via HU-IH, Huey (Iroquois) troop transport helicopters from 1966 to the end of the Vietnam War in 1975.

AH-1 "Cobra" Attack Helicopter: The Bell AH-1 Cobra is a single-engine, two-pilot attack helicopter designed for direct fire support of ground troops. Developed from the Huey transport helicopter, it was the first purpose-built helicopter gunship to enter military service. It is still in service with the U.S. Marine Corps today.

Air Assault School: The United States Army Air Assault School is a two-week, specialized training course that prepares soldiers for combat assault operations involving rotary-wing (helicopter) aircraft.

Airborne School: The United States Army Airborne School, widely known as Jump School, conducts the three-week basic paratrooper (military parachutist) training for the United States Armed Forces. It was founded in 1940 prior to the U.S. entering World War II.

American Battle Monuments Commission (ABMC): AMBC is an independent agency of the United States government that administers, operates, and maintains permanent U.S. military cemeteries, memorials and monuments primarily outside the United States. There were 26 cemeteries and 31 memorials, monuments and markers under the care of the ABMC. There are more than 140,000 U.S. servicemen and servicewomen interred at the cemeteries.

Army Commendation Award (ARCOM): The Army Commendation Medal (ARCOM) is a mid-level award that is granted for consistent acts of heroism or meritorious service. The award is given by a local commander allowing for generous interpretation of the criteria for which the medal is given.

Area of Responsibility (AOR): A pre-defined geographic region designed to allow a single commander to exercise command and control of all military forces in the AOR, regardless of their branch of service.

Armed Forces Staff College: The Armed Forces Staff College, now known as the Joint Forces Staff College (JFSC), is in Norfolk, Virginia. Established on August 13, 1946. Today the JFSC educates and acculturates joint and multinational warfighters to plan and lead at the operational level. Its mission is to prepare national security professionals for planning and executing joint, multinational, and interagency operations.

Armed Forces Vocational Aptitude Battery (AFVAB): The Armed Forces Vocational Aptitude Battery (AFVAB) is a multiple-choice test used to determine qualification for enlistment in the United States Armed Forces. It assesses an individual's aptitude in various areas, predicting their success in military occupations and academic pursuits.

Army Enlistment Contract: The Army Enlistment Contract is a legal document that outlines the terms and conditions of an individual's service in the United States Army. It is a binding agreement between the individual and the Army, specifying the duration of service, job assignment, training, and other obligations.

Army Individual Ready Reserves (IRR): The Individual Ready Reserve (IRR) is a category of the Ready Reserve of the Reserve Component of the Armed Forces of the United States composed of former active duty or reserve military personnel. The IRR is composed of enlisted personnel and officers, with more than 200 Military Occupational Specialties (MOS), are represented, including combat arms, combat support, and combat service support.

Army National Guard: The National Guard is a unique and essential element of the U.S. military. Founded in 1636 as a citizen force organized to protect families and towns from hostile attacks, today's National Guard Soldiers hold civilian jobs or attend college while maintaining their military training part-time, always ready to defend the American way of life in the event of an emergency.

Army Non-Commissioned Officers (NCO) Creed: The Army NCO Creed is a guiding principle for enlisted leaders, emphasizing their responsibilities, authority, and professionalism.

Army Reserve Command (formerly ARCOM): The United States Army

Reserve Command (USARC) is the headquarters command responsible for overseeing all United States Army Reserve units.

Army Reserves (USAR): Is a reserve force of the United States Army. It is a part-time military force composed of citizen soldiers who serve alongside the active-duty Army. The Army Reserve is the largest branch of the US Armed Forces' reserve components, with approximately 188,703 reserve members.

Army Security Agency (ASA): The Army Security Agency (ASA) was the United States Army's signals intelligence branch from 1945 to 1976. Headquartered at Arlington Hall Station, Virginia, ASA was responsible for communications intelligence, electronic intelligence, signals intelligence, and communications security. The National Security Agency absorbed the personnel and assets of ASA in 1976.

Army Special Operations Forces (ARSOF): It is an Army Service Component Command. Its mission is to organize, train, educate, man, equip, fund, administer, mobilize, deploy and sustain Army special operations forces to successfully conduct worldwide special operations.

Army Test and Evaluation Program (ARTEP): The Army Test and Evaluation Program (ARTEP) is a comprehensive training and evaluation framework for Army units, focusing on maintaining operational readiness.

Army War College: Provides graduate-level instruction to senior military officers and civilians to prepare them for senior leadership assignments and responsibilities.

Assistant Division Commander: Assists the division commander in the execution of readiness programs, with special emphasis on strength maintenance, personnel and training readiness. Work involves directly supervising and evaluating brigade commanders.

Association of the United States Army (AUSA): The Association of the United States Army is a nonprofit educational and professional development association serving America's Army and supporters of a strong national defense. AUSA provides a voice for the Army, supports the Soldier, and honors those who have served in order to advance the security of the nation.

Aviation Squadron: An Aviation unit comprising a number of military aircraft and their aircrews, usually of the same type, typically with 12 to 24 aircraft, sometimes divided into three or four flights, depending on aircraft type.

Army Aviator School: Is a military educational establishment responsible for training and developing personnel and equipment for the Army's aviation branch. The school is located at Fort Novosel, Alabama.

Army Basic Combat Training (BCT): US Army Basic Combat Training (BCT) is a 10-week program that transforms civilians into Soldiers, teaching them the skills and values necessary to succeed in the Army.

Army Basic Non-Commissioned Officers Course (BNOC): This was a basic leadership course for Sergeants (E-5) Professional Military Education (PME). Today it is known as Advanced Leader Course (ALC).

Battalion: The word "battalion" came into the English language in the 16th century from the French *bataillon*, meaning "battle squadron." It is a unit formation consisting roughly of 600-1000 persons, led by a Lieutenant Colonel and consists of three "line" companies, a support company and a Headquarters Company, typically led by a Captain.

Battalion Command Sergeant Major: The senior enlisted member in a Battalion and senior enlisted advisor to the Battalion Commander.

Battalion Commander: Senior officer of a Battalion, typically a Lieutenant Colonel.

Battalion Executive Officer: Second most senior officer in a Battalion, typically a Major.

Battalion Staff: An Army battalion staff is a group of officers and enlisted personnel who serve as advisors to the battalion commander, providing expertise in various areas to support the commander's decision-making and execution of tasks. The staff is organized into functional sections, each responsible for a specific area of responsibility. S-1 (Personnel) is responsible for personnel management, including administrative and logistical support. S-2 (Intelligence) focuses on intelligence gathering, analysis, and dissemination to support operational planning and execution. S-3 (Operations) develops and executes operational plans, coordinates training, and manages the battalion's daily activities. S-4 (Logistics) oversees supply, maintenance, and transportation, ensuring the battalion's material readiness. S-5 handles civil-military relations and is in charge of the civil-military relations center. S-6 handles communication systems, networks, and equipment to facilitate information exchange within the battalion and with higher headquarters.

Beginning Morning Nautical Twilight (BMNT): Nautical Dawn. It is used and considered when planning military operations. A military unit may treat BMNT and EENT with heightened security, e.g. by "standing to", in which everyone assumes a 100% alert defensive position.

Board of Cooperative Education Services (BOCES): Provides shared educational programs and services to school districts within New York state.

Brigade: A brigade is a major tactical military formation that typically comprises three to six battalions plus supporting elements. It is roughly equivalent to an enlarged or reinforced regiment. Two or more brigades may constitute a division.

Brigade Combat Team (BCT): A Brigade Combat Team (BCT) is the basic deployable unit of maneuver in the U.S. Army. It consists of one combat arms branch maneuver brigade (e.g., infantry, armored, Stryker) along with its assigned support and fire units. A brigade is typically commanded by a colonel (O-6), although in some cases a brigadier general (O-7) may assume command.

Brigadier General (BG O-7): A brigadier general ranks above a colonel and below a major general. The pay grade of brigadier general is O-7.

C-130 'Hercules' Cargo Aircraft: The C-130 Hercules is a medium-sized, multi-role, multi-mission transport aircraft designed for military and civilian use. It is capable of operating from rough, dirt strips and is the prime transport for airdropping troops and equipment into hostile areas.

C-141 Starlifter: The Lockheed C-141 Starlifter is a retired military strategic airlifter that served with the United States Air Force (USAF) from 1965 to 2006. It was designed to meet military standards as a troop and cargo carrier, and its primary role was to rapidly transport U.S. Army troops and equipment anywhere in the world.

Cadre: A nucleus of trained personnel around which a larger organization can be built and trained. A cadre of Non-Commissioned Officers and Officers who train troops.

Calvary Regiment: A typical cavalry regiment consists of several troops, each with its own commander and personnel. The regiment is usually commanded by a colonel or lieutenant colonel. Cavalry regiments have historically performed various tasks, including reconnaissance and scouting, security and screening, flanking and attacking enemy positions, pursuing and disrupting enemy retreats, supporting infantry and artillery operations.

Captain (CPT O-3): An Army Captain is a commissioned officer rank in the United States Army. It is a company-grade officer rank, typically commanding a company-sized unit of 60-200 soldiers and/or serving on a battalion, brigade and division staff, which may include infantry, artillery, armor, or cavalry units.

Central Command (CENTCOM): The United States Central Command is one of the eleven unified combatant commands of the U.S. Department of Defense. It was established in 1983. Its Area of Responsibility (AOR) includes the Middle East, including Egypt in Africa, Central Asia and parts of South Asia. The command has been the main American presence in many military operations, including the Persian Gulf War's Operation Desert Storm in 1991, the Wars in Afghanistan and Iraq from 2001 to present day.

Chief of Staff of the Army (CSA): The chief of staff of the Army (CSA) is a statutory position in the United States Army held by a general officer. As the highest-ranking officer assigned to serve in the Department of the Army, the chief is the principal military advisor and a deputy to the secretary of the Army. In a separate capacity, the CSA is a member of the Joint Chiefs of Staff and, thereby, a military advisor to the National Security Council, the secretary of defense, and the president of the United States.

Colonel (COL O-6): A US Army Colonel (O-6) is a senior field-grade officer, ranking above Lieutenant Colonel (O-5) and below Brigadier

General (O-7). Colonels typically command a brigade-sized unit consisting of 3,000 to 5,000 soldiers, with the assistance of several subordinate commissioned officers and a Command Sergeant Major as a primary non-commissioned officer advisor.

Combat Diver School: The Army Combat Diver Qualification Course (CDQC) is a rigorous seven-week training program that teaches selected Special Operations Forces (SOF) personnel to conduct combat diving operations.

Combat Infantryman Badge (CIB): The CIB is awarded to Army personnel in the grade of Colonel or below, and warrant officers with infantry or special forces Military Occupational Specialty (MOS). The soldier must actively participate in ground combat while serving in an assigned infantry or special forces primary duty, in a unit engaged in active ground combat with the enemy.

Combatant Command: A unified combatant command, also referred to as a combatant command (CCMD), is a joint military command of the United States Department of Defense that is composed of units from two or more service branches of the United States Armed Forces and conducts broad and continuing missions. There are currently 11 unified combatant commands.

Command Sergeant Major (CSM E-9): The Command Sergeant Major (CSM) is the highest-ranking non-commissioned officer (NCO) in the United States Army, serving as the senior enlisted advisor to unit commanders. The CSM is responsible for providing counsel, guidance, and oversight to ensure the effective leadership and well-being of all soldiers within their unit.

Commandant of Cadets: The Commandant of Cadets is the ranking officer in charge of the Corps of Cadets at the United States Military Academy, and all cadets enrolled in the Reserve Officer Training Corps (ROTC) units. The commandant is responsible for the administration, discipline, and military training of the cadets.

Commissioned Officer: Is a soldier who has been trained to be a leader in the U.S. Army by successfully graduating from the U.S. Military Academy at West Point, NY, a ROTC program, Officer Candidate School or by Direct Commission due to either specialized training, e.g. lawyer, doctor, nurse, or in much more rare circumstances during war. They typically are the officer in charge at all levels of organized units in three tiers, Company Grade Officers (O1: O3), Field Grade Officers (O4: O6) and General Officers (O7: O10).

Company: A US Army Company is a military unit that typically consists of 60 to 200 soldiers, commanded by a captain or major. It is a mid-level unit, situated between the platoon (20-50 soldiers) and the battalion (300-1,000 soldiers). A company in the US Army is normally composed of three to four platoons, each with its own specific function and responsibilities. The company's structure and organization can vary depending on its specific role, such as infantry, armor, artillery, or support units.

Company Commander (CO): Is a Captain (O3) leading a company unit of 60 to 200 soldiers. The company commander oversees daily operations, training, and logistics for their company, provides leadership and guidance to junior officers and non-commissioned officers (NCOs), makes tactical decisions and executes orders from higher headquarters, may serve as a staff officer in regimental or brigade headquarters.

Company Executive Officer (XO): The Company XO is the second-in-command of a US Army company, typically the senior First Lieutenant (O-2). The XO serves as the principal assistant to the Company

Commander, overseeing the day-to-day administrative and operational activities of the company. The XO is an administrative staff position, focusing on behind-the-scenes tasks rather than direct command of soldiers, works closely with the First Sergeant (1SG) to execute the company's vision and manage daily activities to allow the Company Commander handle "Future Operations," focusing on strategic planning and direction.

Company First Sergeant (1SG/MSG): The First Sergeant is the senior ranking NCO in a company, typically an E-8 position. The First Sergeant is considered a leadership rank while a Master Sergeant (equal in pay grade to a First Sergeant) is considered a staff rank. Duties include senior enlisted advisor to the company commander, overseeing daily operations and ensuring the company meets its mission requirements, manages company administrative processes, advises the commander on all enlisted matters, responsible for the health, welfare, training, morale, and professional development of company soldiers, selects unit personnel to perform as support staff and assists the commander in planning and conducting unit training and supervises and mentors senior staff and coordinates activities of all duty positions.

Corporal/Specialist Fourth Class (SPC4 E-4): It is a junior enlisted rank commonly referred to as "Specialist." It is the senior enlisted rank of the junior enlisted layer (E-1 to E-4). May serve as a Fire Team Leader and serves as a technical expert or skilled soldier in their military occupation specialty (MOS) in various units, including combat, support, or service units. Performs specialized tasks and duties, often requiring advanced training or certification.

Corps: A corps is an "operational unit of employment", that may command a flexible number of modular units. Usually commanded by a lieutenant general. 20,000–45,000 soldiers.

Delayed Entry Program (DEP): DEP allows new enlistees to join the Army before shipping out to basic training. DEP accommodates new enlistees who need time to complete tasks such as graduating from high school or college, preparing physically and mentally for military life, and putting personal affairs in order. DEP enlistees may participate in Army fitness programs and receive an orientation program, including a welcome kit, review of commitments, job assignments, and other important details.

Demilitarized Zone (DMZ): A DMZ is a heavily militarized strip of land dividing a country. In an example, the DMZ running across the Korean Peninsula near the 38th parallel north is a border barrier that divides the peninsula roughly in half. It was established to serve as a buffer zone between the countries of North Korea and South Korea under the provisions of the Korean Armistice Agreement in 1953, an agreement between North Korea, China, and the United Nations Command.

Department of the Army Civilian (DAC): A Department of the Army Civilian (DAC) is a civilian employee of the United States Army, serving under the Department of the Army (DA). DACs are not military personnel, but rather professionals from various occupations, working alongside active-duty military personnel to support the Army's mission.

Distinguished Flying Cross (DFC): Is a prestigious military decoration awarded for extraordinary aerial achievement or heroism while flying in active operations against the enemy. Established on July 2, 1926, it is the third-highest military decoration for valor in the United States.

Distinguished Service Cross (DSC): The Distinguished Service Cross (DSC) is a prestigious military decoration awarded for extraordinary heroism in combat. It is the second-highest military decoration in the US Army, behind the Medal of Honor,

District of Columbia (DC): The District of Columbia, commonly referred to as Washington, D.C., is the capital city of the United States.

Division: A large military unit or formation, usually consisting of between 10,000 and 25,000 soldiers. In most armies, a division is composed of several regiments or brigades; in turn, several divisions typically make up a corps. Historically, the division has been the default combined arms unit capable of independent operations. Smaller combined arms units, such as the American regimental combat team (RCT) during World War II, were used when conditions favored them.

Drill Sergeant: A Drill Sergeant/Instructor (DI) is a non-commissioned officer (NCO) responsible for training and indoctrinating new recruits in military service. The term "Drill Sergeant" is used in the United States Army.

Emergency Medical Technician (EMT): An EMT is an entry-level standard of practitioner in the ambulance service, providing out-of-hospital emergency medical care and transportation for critical and emergent patients.

End of Exercise (ENDEX): Announces the administrative end of a military exercise.

Ending Evening Nautical Twilight (EENT): It is the instant of last available daylight for the visual control of limited military operations.

Family Readiness Group (FRG): The Army FRG is a unit-based organization that supports the families of Army soldiers, particularly during deployments. The FRG's primary goal is to maintain family readiness by providing information, resources, and support to help families adapt to the challenges of military life.

Family Support Group (FSG): The original unit-based organization to support Army families. It has been replaced with the Family Readiness Group.

Field Wireman: The MOS of Field Wireman no longer exists. It was a 25 series MOS for the Signal Corps.

Fire Team: An Army Fire Team is a subunit of an Army Squad, consisting of 2-5 soldiers. A Fire Team is a small, autonomous element designed to optimize "NCO initiative," "combined arms," "bounding overwatch," and "fire and movement" tactical doctrine in combat.

Fire Team Leader: A Fire Team Leader (FTL) is a critical leadership position within an infantry squad in the United States Army. They are responsible for commanding a Fire Team, which is the smallest maneuver element in the Army, consisting of three to five soldiers.

First Lieutenant (1LT O-2): A First Lieutenant (1LT) is a commissioned officer in the United States Army, ranking above Second Lieutenant (2LT) and below Captain (CPT). It is a junior officer rank, typically held by officers with 1-3 years of service. A 1LT typically is a Platoon leader commanding a platoon-sized element (15-50 soldiers) in combat or non-combat environments, a staff officer serving in a company, battalion, or higher headquarters, providing operational and administrative support or a specialty officer leading a specialty platoon, such as an engineer, aviation, or medical unit.

Flight Aptitude Selection Test (FAST): The Army Flight Aptitude Selection Test (FAST) is a standardized assessment used to evaluate candidates for Army flight training programs. The test is designed to measure an individual's ability to manage a dynamic environment, think critically, and apply problem-solving skills.

Fort Benning, GA (Today Fort Moore): Fort Moore (formerly **Fort Benning**) is a United States Army post near Columbus, Georgia. Fort Moore is the home of the United States Army Maneuver Center of Excellence, the United States Army Armor School, United States Army Infantry School, the Western Hemisphere Institute for Security Cooperation (formerly known as the School of the Americas), headquarters of the 75th Ranger Regiment, 3rd Ranger Battalion, Ranger School, Airborne School Pathfinder School and other tenant units.

Fort Bragg, NC (Today Fort Liberty): Fort Liberty, formerly **Fort Bragg,** is a military installation of the United States Army in North Carolina and is one of the largest military installations in the world by population, with over 52,000 military personnel. It is the home of the Army's XVIII Airborne Corps and is the headquarters of the United States Army Special Operations Command, which oversees the U.S. Army 1st Special Forces Command (Airborne) and 75th Ranger Regiment.

Fort Leonard Wood, MO: Fort Leonard Wood is a premier Army Center of Excellence, training and educating service members from the United States Army, Air Force, Coast Guard, Navy, and Marine Corps. The base is home to the Maneuver Support Center of Excellence, the U.S. Army Chemical, Biological, Radiological, and Nuclear School, the U.S. Army Engineer School, and the U.S. Army Military Police School.

Fort Lewis, WA: Fort Lewis is a United States Army base located in Pierce County, Washington, approximately 9.1 miles south-southwest of Tacoma. It is one of the largest and most modern military reservations in the United States, spanning 87,000 acres (136 sq mi; 350 km2) of prairie land cut from the glacier-flattened Nisqually Plain. In 2010, Fort Lewis merged with McChord Air Force Base to form Joint Base Lewis-McChord (JBLM), one of 12 joint bases worldwide. It has been the home of the 2nd Ranger Battalion for 50 years.

Fulda Gap, Germany: The Fulda Gap, located in central Germany, refers to a lowland corridor running southwest from the state of Thuringia to Frankfurt am Main. During the Cold War, it was identified by Western strategists as a potential route for a Soviet invasion of the American occupation zone from the eastern sector occupied by the Soviet Union.

General (GEN, O-10): A US Army General is a senior officer rank in the United States Army, typically holding the highest authority within the service branch. The rank is above Lieutenant General and below General of the Army.

GI Bill: The GI Bill originated in 1944 as the Servicemen's Readjustment Act, providing education, unemployment insurance, and housing benefits to World War II veterans. The Post 9/11 GI Bill expanded benefits to include tuition, fee, and housing stipends for eligible veterans and their family members. The GI Bill has undergone revisions and expansions to accommodate changing veterans' needs and remains a vital program for supporting their education and career goals.

Grazing Fire: Grazing fire is a term used in military science, defined by NATO and the United States Department of Defense as "Fire approximately parallel to the ground where the center of the cone of fire does not rise above one meter (~3 ft) above the ground."

GT Score: The GT Score, also known as the General Technical score, is a subsection of the Armed Services Vocational Aptitude Battery (ASVAB) test. It's a composite score calculated from three subjects, Arithmetic Reasoning: tests mathematical problem-solving skills, Paragraph Comprehension: evaluates reading comprehension and understanding of written passages and Vocabulary: assesses knowledge of word meanings and usage. It is used by the Army and Marines to identify talented recruits for various roles and training programs. It's an important line score, as it helps determine eligibility for specific Military Occupational Specialties (MOS).

Honor Graduate: The title recognizes outstanding individual performance during attendance to Army academic and training schools. Typically, in most Army schools there will be three awards for superior individual performance: Distinguished Honor Graduate, Honor Graduate, and Leadership Award.

Independent Order of the Odd Fellows: Is a fraternal, non-profit charity dedicated to improving and elevating the character of mankind by promoting the principles of friendship, love, truth, faith, hope, charity and universal justice. Their mission is to help make the world a better place to live in, by aiding each other, the community, the less fortunate, the youth, the elderly, and the environment in every way possible, by promoting good will and harmony amongst peoples and nations through the principle of universal fraternity, holding the belief that all men and women regardless of race, nationality, religion, social status, gender, rank and station are brothers and sisters.

Joint Multinational Readiness Center (JMRC): The mission of the of the Joint Multinational Readiness Center (JMRC), in a forward deployed environment at Grafenwoehr and Hohenfels, Germany, is to provide tough, realistic, and challenging joint and combined arms training; focuse on improving readiness by developing soldiers, their leaders and units in support of the Global War on Terrorism, and for success on current and future battlefields; provide simulated combat training exercises for task organized Brigade Combat Teams (BCT)/Heavy BCT (HBCT), Stryker BCT (SBCT), Airborne BCT (ABCT), and functional brigades across the full spectrum of operations to prepare units for full spectrum operations: Major Combat Operations (MCO), Counter-Insurgency (COIN) Operations, and Security Operations Stability Operations (SOSO).

Joint National Training Capability (JNTC): The JNTC was assigned to Joint Forces Command (JFCOM) for establishment and administration, with the goal of providing training capabilities for a joint military force. Initially envisioned as a physical center, later redefined as a network of computer simulations.

Joint Readiness Training Center (JRTC): The Joint Readiness Training Center (JRTC) is a United States Army training center located at Ft. Johnson, LA (Formerly Ft. Polk) focused on improving unit readiness by providing highly realistic, stressful, joint and combined arms training across the full spectrum of conflict (current and future). Its mission is to train Soldiers and grow leaders to deploy, fight, and win.

Joint Special Operations Command (JSOC): JSOC is a joint headquarters designed to study special operations requirements and techniques; ensure interoperability and equipment standardization; plan and conduct joint special operations exercises and training; develop joint special operations tactics. JSOC oversees the Special Mission Units of U.S. Special Operations Command. These are ultra-elite special operations forces units that conduct highly classified and complex operations.

Joint Staff: The Joint Staff is a component of the Joint Chiefs of Staff (JCS) within the United States Department of Defense (DoD). It provides staff support to the Chairman and Vice Chairman of the Joint Chiefs of Staff, as well as to the service chiefs and the National Guard Bureau chief.

Joint Staff J-7: The Joint Staff J-7 is responsible for joint force development (JFD) to advance the operational effectiveness of the current and future joint force. J-7 performs its duties across the spectrum of joint force development by focusing on core functions to train, educate, develop, design and adapt the Joint Force.

Jumpmaster: Jumpmasters are the expert paratroopers in an airborne unit who train, teach and conduct jumpmaster duties for jumping from airplanes.

Jumpmaster School: Trains personnel in the skills necessary to jumpmaster a combat-equipped jump and the proper attaching, jumping, and releasing of combat and individual equipment while participating in an actual jump that is proficient in the duties and responsibilities of the Jumpmaster and Safety.

Jungle Operations Training Course (JOTC): JOTC is a two-week course designed to prepare soldiers for jungle operations, focusing on skills such as: jungle survival, communication, navigation, waterborne operations, combat tracking, jungle small unit tactics, situational training exercises at the squad level. It was originally at Ft. Sherman, Republic of Panama and now is located at the Lightning Academy, 25th Infantry Division, Schofield Barracks East Range, Hawaii.

Junior Reserve Officer Training Corps (JROTC): The Junior Reserve Officer Training Corps (JROTC) is a federal program sponsored by the United States Armed Forces in high schools and some middle schools across the United States and at US military bases worldwide. Established in 1916, JROTC aims to develop citizenship, character, and leadership skills in high school students.

Kill Zone: In military tactics, a Kill Zone (also known as a Killing Zone) is an area entirely covered by direct and effective fire, a critical element of an ambush. It is the primary target area where an approaching enemy force is trapped and destroyed.

Korean War: The Korean War was a conflict between North Korea, supported by China, and South Korea, supported by the United Nations with the United States as the principal participant. The war lasted from 1950 to 1953.

Land Navigation: Land Navigation is the discipline of following a route through unfamiliar terrain on foot or by vehicle, using maps with reference to terrain, a compass, and other navigational tools. It is a core military discipline, essential for military training, and has developed into a sport known as orienteering.

Landing Zone (LZ): A LZ is a designated area where aircraft, particularly helicopters, can safely land and take off. This critical zone is essential for various applications, including military operations, medical evacuations, search and rescue missions

Large Alice Rucksack (Ruck): All-Purpose Lightweight Individual Carrying Equipment (ALICE) introduced in 1974 was made up of components for two types of loads: the "Fighting Load" and the "Existence Load." The ALICE Pack system was designed for use in all environments, whether hot, temperate, cold-wet or even cold-dry arctic conditions.

Lieutenant Colonel (LTC O-5): Is a Field Grade Officer rank in the Army, between the ranks of Major (O4) and Colonel (O6). Typically serves as a Battalion Commander or Executive Officer/Staff Officer in a variety of high-level units or command posts.

Lieutenant General (LTG O-9): A Lieutenant General (LTG) is a commissioned officer in the United States Army, ranking immediately below General (O-10) and above Major General (O-8). The rank is denoted by a three-star insignia. Typically commands an Army Corps, consisting of 60,000 to 70,000 soldiers.

Light Infantry: Light infantry is trained to fight as small groups or individuals, utilizing cover and maneuver to outflank and harass the enemy. Regular infantry, on the other hand, is often organized into massed formations and trained for close combat. Light infantry is designed for rapid movement and flexibility, whereas regular infantry may be slower and more cumbersome due to their heavier equipment. Light Infantry carries lighter weapons and minimal field equipment, whereas regular infantry is equipped with heavier weapons and more extensive gear. Light infantry typically provides a skirmishing screen ahead of the main body of infantry, delaying and harassing the enemy advance. Regular infantry, by contrast, is often the core of large battles, fighting in tight formations.

Load Bearing Equipment (LBE): Load Bearing Equipment (LBE) is a system designed for soldiers to carry combat gear around the waist, utilizing suspenders to help balance the load. It is a crucial component of military load carriage, allowing soldiers to transport essential equipment, ammunition, and supplies while maintaining mobility and comfort.

Long Range Reconnaissance Patrol (LRRP): The Army Long Range Reconnaissance Patrol (LRRP) is a specialized unit that conducts deep reconnaissance and surveillance operations behind enemy lines. LRRPs are small, heavily armed teams, typically 4-12 men, that patrol extensively in enemy-held territory, gathering vital intelligence and disrupting enemy command and control structures.

L-Shaped Ambush: Is a tactic used in military operations to conduct ambushes. It involves deploying a group of soldiers in an L-shaped pattern, with one leg of the L parallel to the enemy's expected route of movement (the "kill zone"), and the other leg at a 90-degree angle, providing enfilading fire along the sides of the kill zone.

M16 Rifle: The M16 rifle is a lightweight, air-cooled, gas-operated, magazine-fed, shoulder-fired weapon designed for both automatic and semi-automatic fire. Developed in the late 1950s chambered in 5.56x45mm.

M203 Grenade Launcher: The M203 is a single-shot, 40 mm under-barrel grenade launcher designed to attach underneath the barrel of an M16. Effective firing range is 382 yards (350 meters) for fire-team sized area targets, 164 yards (150 meters) for vehicle or weapon point targets.

M60 Medium Machine Gun: Is a general-purpose machine gun firing 7.62x51mm NATO cartridges from a disintegrating belt. It is a gas-operated, air-cooled, belt-fed, automatic machine gun. The M60 was adopted by the US Army in 1957 and issued to units beginning in 1959. It has served with every branch of the US military and still serves with other armed forces but has been replaced by the M240 machine gun in the early 2000s.

Major (MAJ O-4): A major in the U.S. Army typically serves as a battalion executive officer (XO) or as the battalion operations officer (S3). Majors can also serve as Company Commanding Officers, a major can also serve as a primary staff officer for a regiment, brigade or task force in the areas concerning personnel, logistics, intelligence, and operations. A major will also be a staff officer / action officer on higher staffs and headquarters.

Major General (MG, O-8): A Major General commands a division-sized unit of 10,000 to 17,000 soldiers. The rank is above Brigadier General (O-7) and below Lieutenant General (O-9) in the Army's rank hierarchy.

Master Sergeant (MSG E-8): A Master Sergeant is a senior non-commissioned officer (NCO) who serves as a staff NCO, equal in paygrade to a First Sergeant but with less leadership responsibilities than one. A Master Sergeant often specializes in certain fields or subject matter. In Special Forces, Master Sergeants are the senior NCO in charge of an Operational

Detachment Alpha (A-Team) better known as a Team Sergeant and also serve as staff NCOs on Battalion and Group Staffs.

Mechanized Infantry: Mechanized infantry is a type of ground force that combines infantry troops with armored vehicles, providing a balance between mobility, firepower, and dismounted infantry capabilities. These units are designed to operate effectively in a variety of environments and scenarios, from conventional warfare to peacekeeping and counterinsurgency operations.

Medal of Honor (MOH): The Medal of Honor (MOH) is the United States Armed Forces' highest military decoration, awarded to recognize American soldiers, sailors, marines, airmen, guardians, and coast guardsmen who have distinguished themselves by acts of valor. It is normally awarded by the President, in the name of Congress, to members of the armed forces who have demonstrated conspicuous gallantry and intrepidity at the risk of life above and beyond the call of duty and bravery, courage, sacrifice, and integrity in the face of danger.

Medical Evacuation (MEDIVAC): Medical Evacuation (MEDIVAC) is the timely and efficient movement and enroute care provided by medical personnel to wounded individuals being evacuated from a battlefield to a receiving medical facility. This process involves the use of medically equipped air ambulances, helicopters, and other emergency transport means, including ground ambulances and maritime transfers.

Military Assistance Command Vietnam: Studies and Observations Group (MACV-SOG): MACV-SOG was a top-secret, joint unconventional warfare task force created on January 24, 1964 and operated until 1972. It was a subsidiary command of the Military Assistance Command, Vietnam (MACV). It was a brigade-sized unit, consisting of volunteers from various US military branches, primarily Army Special Forces, Navy

SEALs and Marine Recon NCOs, who conducted clandestine operations in Laos, Cambodia and North Vietnam.

Military District of Washington (MDW): The Military District of Washington (MDW) is a major command of the United States Army, headquartered at Fort Lesley J. McNair in Washington, D.C. Ceremonial support for national events, such as state funerals, including those of former presidents, and ceremonial events at the White House, Pentagon, and Arlington National Cemetery, guarding the Tomb of the Unknown Soldier, participating in joint military ceremonies and events with other branches of the military and government agencies and providing military support to the National Capital Region, including security and logistics.

Military Free Fall (MFF) School: A Special Operations school designed to teach parachute infiltration behind enemy lines from very high altitudes (as high as 25,000' above ground level). The course focuses on vertical wind tunnel body stabilization training, parachute packing, and introduction to military free-fall operations, with a total of 30 military free-fall operations, both day and night, with combat equipment.

Military Intelligence (MI): Military Intelligence (MI) refers to the collection, analysis, and dissemination of information and data to support military operations, planning, and decision-making. It is a critical component of modern warfare, enabling commanders to understand the enemy's capabilities, intentions, and vulnerabilities, and to make informed decisions about tactics, strategy, and resource allocation.

Military Occupational Skill (MOS): MOS is a system used by the United States military to categorize career fields and identify specific jobs. Each MOS represents a unique role or job within the military, covering a wide range of skillsets and levels of responsibility.

Military Operations in Urban Terrain (MOUT): MOUT refers to military actions conducted in urban environments, where human-made construction and infrastructure significantly impact tactical options available to commanders. MOUT involves fighting in densely populated areas, such as cities, towns, and villages, where buildings, streets, and other structures influence the nature of combat.

Multinational Security Transition Command-Iraq (MNSTC-I): MNSTC-I was a training and organizational-support command of the United States Department of Defense, established in June 2004. It was a military formation responsible for developing, organizing, training, equipping, and sustaining the Iraqi Ministry of Defense (MoD) and Ministry of Interior (Iraq) with their respective forces.

National Capital Region (NCR): The NCR is a defined area that encompasses Washington, D.C., and surrounding counties in Maryland and Virginia. The NCR covers approximately 2,500 square miles (6,475 sq km) and is home to over 6 million people. It is a unique region, as it is a collection of sovereign jurisdictions, including federal, state, and local governments, as well as private institutions and communities.

National Ranger Association: The National Ranger Association, Inc. (501C(3)) is a non-profit organization established in 1998. Governed by a board of six members, its primary objectives are to recognize distinguished members in the Ranger community and to promote the values established in the Ranger Creed. The Association supports several programs, including Best Ranger Competition: A prestigious competition recognizing exceptional Rangers, Ranger Hall of Fame: Honoring distinguished Rangers for their outstanding service and achievements, Distinguished Member of the Airborne and Ranger Training Brigade: Recognizing exceptional service and dedication to the Airborne and Ranger Training Brigade

National Security Agency (NSA): The NSA is an intelligence agency of the United States Department of Defense, under the authority of the Director of National Intelligence (DNI). The NSA is responsible for global monitoring, collection, and processing of information and data for foreign intelligence and counterintelligence purposes, specializing in a discipline known as signals intelligence (SIGINT).

NATO Training Mission-Iraq (NTM-I): The NTM-I was a non-combat advisory and capacity-building mission established in 2004 at the request of the Iraqi Interim Government, under the provisions of UN Security Council Resolution 1546. Its primary goal was to assist Iraq in building effective and accountable security forces.

NCO Promotion Board: The Army NCO Promotion Board is a formal evaluation process used to assess the qualifications and readiness of non-commissioned officers (NCOs) for promotion to the next higher rank. The board reviews and scores individual NCOs based on their performance, experience, and demonstrated leadership skills.

Non-Commissioned Officer (NCO): A NCO is a military officer who has risen through the enlisted ranks, typically from Private to Sergeant Major, without receiving a commission as an officer. NCOs are experienced, skilled, and trained professionals who have demonstrated leadership abilities and are responsible for supervising and mentoring junior enlisted soldiers.

North Atlantic Treaty Organization (NATO): NATO was founded after WWII to promote democratic values and enable members to consult and cooperate on defense and security-related issues to solve problems, build trust and, in the long run, prevent conflict. NATO is committed to the peaceful resolution of disputes. If diplomatic efforts fail, it has the military power to undertake crisis-management operations. These are carried out under the collective defense clause of NATO's founding treaty: Article 5 of the Washington Treaty or under a United Nations mandate, alone or in cooperation with other countries and international organizations.

Officer Basic Course (OBC): OBC is a training program designed to develop new combat-effective officers and train them to perform their wartime duties as commissioned officers. It is the second phase of the Basic Officer Leaders Course (BOLC), which is a two-phased training course for commissioned officers in the United States Army. During OBC, officers learn the specifics of their branches, systems, and equipment they will use in their duty units. The course is tailored to each individual branch, e.g., Infantry, Armor, Artillery, etc.

Officer Candidate School (OCS): OCS is a training program that prepares civilians and enlisted personnel to become commissioned officers in the United States Army, Army Reserve, and Army National Guard. OCS trains, assesses, and evaluates potential officers, focusing on leadership skills, military culture, and law. OCS is one of several ways to become a U.S. Army commissioned officer, alongside graduation from the United States Military Academy (USMA), Reserve Officers' Training Corps (ROTC), and Officer Candidate School programs of the Army National Guard.

OH-58 Kiowa Scout Helicopter: The Bell OH-58 Kiowa is a family of single-engine, single-rotor military helicopters used for observation, utility, and direct fire support. Manufactured by Bell Helicopter, the OH-58 was based on the Model 206A Jet Ranger civilian helicopter and has been in continuous use by the United States Army since 1969.

Olive Drab (OD): A specific shade of dark olive green, often referred to as a military color. It is characterized by a matte finish, greyish green color, similar to the hue of a ripe green olive, used frequently for military uniforms and equipment and for camouflage in foliaged environments, particularly in tactical gear and clothing.

On the Job Training (OJT): OJT is a type of instruction that takes place at the workplace, where employees learn new skills and competencies by performing real tasks and responsibilities under the supervision of a trainer, manager, or experienced coworker.

Operation Desert Storm: Operation Desert Storm was a military operation launched by a United States-led coalition in response to Iraq's invasion of Kuwait on August 2, 1990. The operation aimed to expel Iraqi forces from Kuwait and restore the country's sovereignty.

Operation Gothic Serpent: Was a military operation conducted by the United States, supported by the United Nations, in Mogadishu, Somalia, from August to October 1993. The primary objective was to capture Mohamed Farrah Aidid, a Somali military officer and leader of the Somali National Alliance (SNA), who was wanted for attacks against United Nations troops.

Operation Just Cause: Was a military operation conducted by the United States on December 20, 1989, with the primary goal of removing Panamanian General Manuel Noriega from power and capturing him for trial in the United States on drug trafficking and related charges. The operation involved a coalition of US military forces, including the Army, Navy, Marines, Air Force, and Coast Guard, with approximately 27,000 troops participating.

Operation Provide Comfort: Was a humanitarian mission conducted by the United States and its coalition partners in northern Iraq in 1991, following the end of the Gulf War. The operation aimed to provide aid and protection to Kurdish refugees fleeing their homes in the aftermath of the war.

Operation Restore Hope: A multinational humanitarian relief effort led by the United States, sanctioned by the United Nations, and conducted from December 5, 1992, to May 4, 1993, in Somalia. The operation aimed to establish a secure environment for humanitarian relief operations in Somalia, which was torn apart by civil war and famine.

Operation Silver Anvil: Was a non-combatant evacuation operation carried out by United States armed forces in Sierra Leone in April–May 1992. The operation successfully evacuated more than 400 people from the country.

Operational Detachment Alpha (ODA or A- Team): An ODA, also known as an SFOD-A or "A-Team", is the primary fighting force of the Green Berets. It is a specialized tactical team consisting of 12 personnel, led by an 18A (Detachment Commander), a captain, and a 180A (Assistant Detachment Commander), usually a Warrant Officer One or Chief Warrant Officer Two and 10 highly trained NCOs specializing in operations and intelligence, weapons, engineering, medical, and communications.

Order of Merit List (OML): The OML is a ranking system used to evaluate and select non-commissioned officers (NCOs) for promotion, schools, and training. It is a merit-based list, where NCOs are assessed based on their performance, qualifications, and potential for future assignment and leadership.

Panamá Canal: Is an artificial 82-kilometer (51-mile) waterway connecting the Atlantic Ocean with the Pacific Ocean, crossing the Isthmus of Panamá. It is a crucial conduit for maritime trade, allowing ships to bypass the lengthy and hazardous route around South America's southernmost tip.

Panamá Canal Zone: Was a 10-mile-wide (16 km) strip of land along the Panamá Canal, extending from the Atlantic to the Pacific Ocean, and bisecting the Isthmus of Panamá. It was a unique administrative entity, under the jurisdiction of the United States from 1903 to 1979.

Panamá Canal Company: The Panamá Canal Company refers to the entity that operated the Panama Canal from 1904 to 1979. During this period, the company was responsible for the construction, maintenance, and operation of the canal.

Parachute Infantry Regiment: A specialized infantry unit in the United States Army, trained and equipped for airborne operations. These regiments are designed to deploy by parachute, landing behind enemy lines to conduct rapid assaults, secure key objectives, and disrupt enemy communications and supply chains.

Parade Rest: Parade Rest is a formal position of rest assumed by military members. This position is typically executed from the position of attention, and individuals remain silent and motionless unless otherwise directed. Parade Rest is a standardized military posture used in various contexts, including ceremonial events, drills, and other formal situations.

Pathfinder School: The United States Army Pathfinder Course trains military personnel in the U.S. Army and its sister services to set up parachute drop zones and helicopter landing zones for airborne and air assault missions, resupply drops and medivac operations.

Physical Training (PT): PT is a comprehensive program designed to maintain the physical fitness and readiness of Soldiers.

Plank Holder: Is an individual who was a member of the crew of a United States Navy ship or United States Coast Guard Cutter (or Royal Canadian Navy ship) when that ship was placed in commission. Other services have used the term to denote those soldiers, marines or airmen who were original members of newly activated units.

Platoon: Is a military unit typically composed of two to four squads or sections. Platoon organization varies depending on the type of unit, generally can be composed of 20–50 troops. A platoon is typically the smallest military unit led by a commissioned officer, typically a second (O-1) or first lieutenant (O2) or an equivalent rank. The officer is usually assisted by a platoon sergeant who is a Sergeant First Class (E-7).

Platoon Leader: A platoon leader is a junior officer in charge of a platoon, a subunit of a company-sized unit in the army. Typically, a platoon leader is a second or first lieutenant, or an equivalent rank.

Platoon Sergeant (PSG): A noncommissioned officer (NCO) in the grade of E7 who serves as the second-in-command and primary assistant to the platoon leader in the U.S. Army. They are responsible for the platoon's leadership, discipline, training, and welfare, and are expected to set an example for the soldiers.

Pointe du Hoc: A promontory with a 35-meter (110 ft) cliff overlooking the English Channel on the northwestern coast of Normandy, France. During World War II, it was a strategic location fortified by the Germans with a series of bunkers, machine gun posts, and artillery emplacements. On D-Day (June 6, 1944), the 2nd Ranger Battalion, led by Lieutenant Colonel James Earl Rudder, scaled the cliffs using ropes and ladders under heavy enemy fire. Despite initial setbacks and the discovery that the guns had been removed, the Rangers destroyed the firing mechanisms of five of the six guns and neutralized the battery.

Post Traumatic Stress (PTS): PTS is a normal and adaptive response to a traumatic or stressful event. It is a common reaction to experiencing or witnessing a frightening, stressful, or traumatic situation. During and after such an event, the body's "fight-or-flight" response is triggered, preparing the individual to deal with the threat or fear.

Primary Leadership Development Course (PLDC): Was the first course of study in the US Army Noncommissioned Officer Professional Development System (NCOPDS). It is now called the Basic Leadership Course and is a month-long program that trains specialists and corporals in the fundamentals of leadership.

Private (E-1, E-2 and E3): An Army Private is the lowest enlisted rank in the United States Army, with a military paygrade of E-1. Private Second Class is an E-2. And Private First Class is an E-3. These are entry-level ranks for new recruits, typically held during Basic Combat Training (BCT) and their advanced skills training and airborne school. Dependent upon their performance during initial training, many Privates are promoted to E-2 upon completion of BCT and E-3 upon completion of their advanced skills training.

Quiet Professional: A soldier who prioritizes the mission, is humble, selfless, and hardworking, and focuses on delivering results without seeking recognition or fanfare.

Ranger Challenge: The ROTC Ranger Challenge is an extracurricular competition and training program within Army ROTC (Reserve Officers' Training Corps) units. It is designed to test cadets' mental and physical toughness, leadership, teamwork, and esprit-de-corps. A two-day event conducted on a non-tactical course, featuring nine graded events focusing on developing technical competence, leadership, teamwork, and physical endurance.

Ranger Creed: The Ranger Creed is a guiding philosophy and code of conduct for the 75th Ranger Regiment, written by Command Sergeant Major Neal R. Gentry in 1974. It embodies the spirit, discipline, and duty expected of Rangers in both peace and war.

Ranger Enlisted Option: From 1970 until 1975, the Ranger Enlisted Option allowed new Army recruits to attend Ranger School after completing Basic, AIT and Airborne School. Because most of these new recruits were still privates and had very little tactical experience, there was a very high attrition rate. The option was changed in 1975 to allow recruits to be assigned to a Ranger Battalion after Airborne School. Today, they call it Option 40. It allows recruits to attend the Ranger Assessment and Selection Program (RASP) after completing Airborne School. This is a very difficult assessment program, and individuals must pass the course meeting or exceeding all standards. It is required to be assigned to the Ranger Regiment. All volunteers, including officers and NCOs must attend RASP as well.

Ranger Regiment: The 75th Ranger Regiment, also known as the Army Rangers, is the premier light infantry and direct-action raid force of the United States Army Special Operations Command. The 75th Ranger Regiment is one of the U.S. military's most extensively used units. On December 17, 2020, it marked 7,000 consecutive days of combat operations.

Ranger School: Ranger School is a 62-day United States Army small unit tactics and leadership course that develops functional skills directly related to units whose mission is to engage the enemy in close combat and direct fire battles. Ranger training was established in September 1950 at Fort Benning, Georgia (now called Fort Moore). The Ranger course has changed little since its inception. It is an eight-week course divided into three phases. Benning Phase, Mountain Phase, and Swamp Phase.

Ranger Training Brigade: The Ranger Department was established in October 1951, and the first Ranger class for individual candidates graduated on March 1, 1952. In 1987, the Ranger Department reorganized into the Ranger Training Brigade, establishing four Ranger Training Battalions. In 2014, the brigade changed its name to the Airborne and Ranger Training Brigade, consolidating its training programs under one central location.

RB 15 Rubber Boat: The RB 15 Rubber Boat is a 15-passenger inflatable boat used by military forces, particularly by special operations units and commandos. It is designed for covert operations, reconnaissance, and insertion/extraction missions.

Regiment: A US Army regiment is a military unit composed of troops headed by a colonel and organized for tactical control into companies, battalions, or squadrons. The term "regiment" originates from the Latin "regimen," meaning "rule" or "system of order," describing the unit's functions of raising, equipping, and training troops. It consists of 1000-1500 troops.

Replacement Company (Depot): The Army Replacement Company is a unit responsible for providing replacements for combat divisions in the regular army. It is a critical component of the army's personnel system, ensuring that front-line formations maintain high numerical strength during prolonged combat.

Reserve Officer Training Corps (ROTC): ROTC is a group of college- and university-based officer-training programs for training commissioned officers of the United States Armed Forces. While ROTC graduate officers serve in all branches of the U.S. military, each branch has its own distinct program.

"Retired on Active Duty" (ROAD): Is a derogatory term used typically to describe an aging senior NCO or Officer who is close to retirement and has elected to sit back and do as little as possible until they are officially retired.

Rotary Wing: A rotorcraft or rotary-wing aircraft is a "heavier-than-air" aircraft with rotary wings or rotor blades, which generate lift by rotating around a vertical mast. Several rotor blades mounted on a single mast are referred to as a rotor. Helicopters are Rotary Wing Aircraft.

Scout Platoon: Is a specialized military unit that conducts reconnaissance and security operations to support its parent unit. It serves as the commander's "eyes and ears" on the battlefield, employing stealth and proper techniques of movement to gather information about the enemy and terrain.

Scout Swimmer School: Scout Swimmer school is a specialized training program that equips Marines and soldiers with amphibious capabilities, enabling them to conduct beach reconnaissance and secure landing sites ahead of a main raiding force. The program focuses on teaching Marines to swim, navigate, and operate in various water environments, while carrying combat gear and equipment.

Second Lieutenant (2LT O-1): It is a junior officer rank, typically assigned to newly commissioned officers who have completed their officer training.

Sergeant (SGT E-5): An E5 Sergeant is a non-commissioned officer (NCO) rank, also known as a junior NCO. It is the first NCO rank, above the Specialist/Corporal (E4) and below the Staff Sergeant (E6).

Sergeant First Class (SFC E-7): A Sergeant First Class (SFC E-7) is the seventh enlisted rank in the United States Army, ranking above Staff Sergeant (E-6) and below Master Sergeant and First Sergeant (E-8). It is the first non-commissioned officer (NCO) rank designated as a senior non-commissioned officer (SNCO).

Sergeant Major (SGM E-9): A Sergeant Major (SGM) is the highest enlisted rank, with a pay grade of E-9. It is a senior non-commissioned officer (NCO) position, and the SGM serves as a leader, advisor, and representative of the Army's enlisted force.

Services/Military Departments: These are the independent branches of the military that fall under the Secretary of Defense: Army, Navy, Airforce, Marine, Coast Guard and Space Force. Each are led by a Service Secretary and a senior, 4-star Chief of Staff or Naval Operations.

Silver Wings: Slang for the US Army Parachute Badge, also known as the Jump Wings. It is the military badge awarded to soldiers who have completed the US Army Airborne School and are qualified military parachutists. The badge is silver and signifies that the soldier is trained and qualified to participate in airborne operations.

Sniper School: A specialized training program located at Fort Moore, Georgia, within the 199th Infantry Brigade, U.S. Army Infantry School. Established in 1986, it is one of the longest-running sniper training courses in the US Army's history. The seven-week course is designed to produce highly skilled snipers who can operate effectively in a variety of environments.

Special Forces (SF: Green Berets): The United States Army Special Forces, colloquially known as the "Green Berets," is one of several special operations units in the United States Army. This elite unit focuses on four functional areas, unconventional warfare, foreign internal defense, direct action and special reconnaissance. They have extensive foreign language training and cultural expertise about their assigned Area of Operation.

Special Forces Qualification Course (SFQC): The US Army Special Forces Qualification Course (SFQC), also informally referred to as the "Q Course," is the initial formal training program for entry into the United States Army Special Forces. The course consists of four phases: Phase I: Special Forces Assessment and Selection (SFAS): A rigorous assessment and selection process to determine a candidate's suitability for advanced Special Forces training. Phase II: Basic Skills Phase: Focuses on developing fundamental skills in areas such as language, first aid, combat tactics, and survival techniques. Phase III: Advanced Skills Phase: Builds on the basic skills, emphasizing specialized training in areas like unconventional warfare, foreign internal defense, and direct action. Phase IV: Collective Training Phase: A realistic Unconventional Warfare (UW) culmination exercise where soldiers apply their skills in a simulated operational environment.

Special Operations Command (SOCOM): The United States Special Operations Command (USSOCOM or SOCOM) is the unified combatant command charged with overseeing the various special operations component commands of the Army, Marine Corps, Navy, and Air Force of the United States Armed Forces. The command is part of the Department of Defense and is the only unified combatant command created by an Act of Congress. USSOCOM is headquartered at MacDill Air Force Base in Tampa, Florida.

Special Operations Forces (SOF): Are highly trained and specialized, elite military units within the United States and NATO militaries. They are designed to conduct complex, dynamic, and often clandestine operations in various environments.

Special Weapons and Tactics (SWAT): SWAT teams are specialized units within law enforcement agencies, trained to respond to high-risk situations that require specialized skills, tactics, and equipment. These teams are designed to neutralize threats, protect public safety, and minimize harm to all parties involved.

Squad: An army squad is the smallest organizational grouping of personnel in a military unit, typically commanded by a Staff Sergeant (E-6) squad leader. The squad consists of nine soldiers broken into two fire teams of four soldiers, led by a Sergent (E-5) and is designed to perform specific combat or support tasks.

Squad Leader: A squad leader is typically a Staff Sergeant (E-6) or a Sergeant (E-5).

Staff Sergeant (SSG E-6): A SSG is above Sergeant (E-5) and below Sergeant First Class (E-7) in the Army's enlisted rank hierarchy. A Staff Sergeant typically commands a squad of nine to ten soldiers, sometimes more, and is responsible for leading, training, and mentoring their team.

Standard Operating Procedures (SOP): A SOP is a set of step-by-step instructions that outline the specific tasks, methods, and protocols to be followed in various situations. These procedures are unique to each unit and are not necessarily standard across all units. In the US Army, SOPs are used to ensure consistency and efficiency in operations, particularly when units are operating independently or without direct command guidance.

Straight Leg (Leg): This slight dates to WWII. A "Straight Leg" refers to a non-Airborne soldier, as opposed to an Airborne soldier. Also known as a "Leg." The term "Straight Leg" originates from the historical practice of non-Airborne soldiers not bending their legs at the knee in preparation for landing, unlike Airborne troops who do so when jumping from aircraft.

Strategic Headquarters Allied Powers Europe (SHAPE): SHAPE is the Supreme Headquarters of the Allied Powers in Europe, serving as the military headquarters of the North Atlantic Treaty Organization's (NATO) Allied Command Operations (ACO). It is responsible for planning and executing all NATO operations worldwide.

Stryker Combat Vehicle: The Stryker is a family of eight-wheeled armored combat vehicles manufactured by General Dynamics Land Systems (GDLS) for the United States Army. It is a hybrid of an infantry fighting vehicle and an armored personnel carrier, designed to provide firepower, battlefield mobility, survivability, and versatility while reducing logistics requirements.

Table of Organization and Equipment (TO&E): Is a document published by the U.S. Army prescribing the organization, staffing, and equipping of Army units based on the type of unit and its mission. It outlines the specific structure, personnel, and equipment required for a unit to perform its intended tasks. This includes organizational structure: The chain of command, positions, and roles within the unit, manpower: The number and types of personnel assigned to the unit and equipment: The specific vehicles, weapons, communications gear, and other assets required for the unit's operations.

Tactical Officer (TAC): TAC Officer and NCOs can be found in various training environments in the Army. This includes the legal Company Commander of a Cadet Company at the United States Military Academy (USMA), West Point, New York, NCO Academies, various leadership schools like BLC, Airborne and Ranger Schools. Their role is to assess the potential and oversee the holistic development of trainees, ensuring they meet the physical, military, academic, and moral-ethical program requirements.

Tactics Instructor: A military professional responsible for teaching and training soldiers, officers, and other personnel in the application of military tactics, techniques, and procedures (TTPs). a military professional responsible for teaching and training soldiers, officers, and other personnel in the application of military tactics, techniques, and procedures (TTPs).

Training and Doctrine Command (TRADOC): TRADOC is a major command of the United States Army, headquartered at Fort Eustis, Virginia. Established on July 1, 1973, its primary role is to oversee the training of Army forces and develop operational doctrine. Its primary responsibilities are to Recruit, train, and educate soldiers, operate 37 schools and centers at 27 different locations, develop and refine operational doctrine for the Army and provide training and guidance to Army units and personnel.

Traumatic Brain Injury (TBI): Combat-related TBI occurs when a service member or veteran sustains a traumatic brain injury (TBI) from an external blow or jolt to the head, typically caused by external forces, such as explosions, blasts, or falls, during military service. It can result in focal or diffuse damage to the brain and may lead to temporary or long-term cognitive, emotional, and behavioral changes. It often co-occurs with post-traumatic stress (PTS).

U.S. Air Force Academy: A four-year federal service academy located just north of Colorado Springs, CO. It is a premier institution for training commissioned officers for the United States Air Force and United States Space Force.

U.S. Army Command and General Staff College (CGSC): CGSC is a graduate school that educates, trains and develops leaders for Unified Land Operations in a joint, interagency, intergovernmental, and multinational operational environment; and to advance the art and science of the Profession of Arms in support of Army operational requirements.

U.S. Army Military Police Corps: Is a branch of the United States Army responsible for policing, detainment, and stability operations.

U.S. Army Ranger Association (USARA): A social membership organization dedicated to supporting active duty and veteran U.S. Army Rangers. Its purpose is to strengthen relationships among all U.S. Army Rangers: past, present, and future; foster camaraderie among those who have earned the title U.S. Army Ranger; and provide an extended community for all US Army Rangers and their families.

U.S. Army Southern European Task Force, Africa (SETAF-AF): Is a formation of the United States Army headquartered at Caserma Ederle, Italy. It is responsible for achieving U.S. Africa Command and U.S. Army Campaign Plan objectives while conducting all U.S. Army operations, exercises, and security cooperation on the African continent.

U.S. Army Special Operations Command (USASOC): USASOC is a major Army command responsible for overseeing the various special operations forces (ARSOF) of the United States Army. USASOC is the largest component of the United States Special Operations Command (USSOCOM). The mission of USASOC is to "organize, train, educate, man, equip, fund, administer, mobilize, deploy and sustain Army special operations forces to conduct successful worldwide special operations."

U.S. Army's School for Advanced Military Studies (SAMS) Fellowship Program: A prestigious educational opportunity for select Armed Forces, Interagency, and Allied members. The program aims to develop critical and creative thinkers, agile and adaptive leaders, and skilled practitioners in doctrine and operational art. It is designed to fill a gap in US military education between the Command and General Staff College's focus on tactics and the War College's focus on grand strategy and national security policy.

U.S. Military Academy (West Point): Is a public liberal arts college located in West Point, New York. Established on March 16, 1802, it is the oldest service academy in the United States and one of the most prestigious institutions for higher education. The USMA provides a four-year program that combines academic excellence with military training, preparing cadets to become commissioned officers in the United States Army.

U.S. Naval Academy (Annapolis): Established on October 10, 1845, it is the second oldest of the five U.S. service academies and the undergraduate college of the naval service. USNA develops midshipmen "morally, mentally, and physically" to become professional officers in the U.S. Navy and Marine Corps.

UH-1H "Huey" Helicopter: The "Huey" is a light-lift utility helicopter designed and built by Bell Helicopter Textron. It is a variant of the UH-1 Iroquois series, commonly known as the "Huey." The UH-1H was produced from 1966 to 1976, with over 7,000 units manufactured. It was the main troop lift helicopter in the Vietnam War.

US Joint Forces Command (JFCOM): JFCOM was a Unified Combatant Command of the United States Department of Defense (DoD), responsible for providing joint warfighting capabilities to combatant commanders. It was established in 1999 and disestablished In 2010 as part of a cost-cutting measure.

USS Pueblo: On January 23, 1968, while conducting a routine surveillance mission, the USS Pueblo was intercepted by North Korean patrol boats approximately 16 miles from the North Korean coast. Despite being in international waters, the North Koreans opened fire and boarded the ship, capturing its 82 crew members and killing one. After 11 months of diplomacy and an apology from the U.S., the crew was released having endured torture and starvation for the period of their captivity.

"Wait-a-minute vine": Names for a variety of prickly plants that catch onto the clothing of soldiers moving through thick underbrush forcing the hooked person to stop ("wait a minute") to remove the thorns carefully to avoid injury or shredded clothing.

War on Terrorism: Also known as the Global War on Terrorism (GWOT), is was an international military campaign launched by the United States following the September 11 attacks in 2001. The campaign aimed to destroy al-Qaeda and other militant extremist organizations that engage in terrorism.

Warrant Officer (WO): A Warrant Officer is a commissioned officer in the armed forces, ranking above non-commissioned officers (NCOs) and below commissioned officers. Warrant Officers are typically technical experts in their field, with a deep understanding of their specialty. They are commissioned officers, taking the same oath as commissioned officers, and are responsible for providing critical perspective and leadership.

Warrant Officer Candidate School (WOCS): WOCS is a five-week course designed to train, assess, evaluate, and develop warrant officers for fourteen of the U.S. Army's sixteen basic branches (excluding Infantry, Armor and Special Forces). The course is designed to provide a base to assist in the development of Army Warrant Officers into self–aware and adaptive technical experts, combat leaders, trainers, mentors, and advisors to both soldiers and commanders. Special Forces Warrant Officers attend the Special Forces Warrant Officer Technical and Tactical Certification Course (SF-WOTTC) at Fort Liberty, North Carolina.

Weapons Platoon: A Weapons Platoon is part of a Light Infantry battalion and is deployed alongside three Rifle Platoons. The Platoon is responsible for providing covering fire and suppressing enemy positions, allowing the Rifle Platoons to advance. It is typically composed of two sections, a Mortar Section equipped with mortars, which are crew-served weapons that fire explosive rounds over long distances to suppress or destroy enemy positions and an Anti-Tank Section armed with anti-tank weapons, such as recoilless rifles or missile launchers, designed to engage and destroy enemy armored vehicles.

Women's Army Corps (WAC): The Women's Army Corps (WAC) was the women's branch of the United States Army, established during World War II. It was created as an auxiliary unit, the Women's Army Auxiliary Corps (WAAC), on May 15, 1942, and converted to an active-duty status in the Army of the United States as the WAC on July 1, 1943. The WAC was disbanded on October 20, 1978, as the Army integrated women into regular units. The Women's Armed Services Integration Act of 1948 allowed women to serve as permanent, regular members of the Army, Navy, Marine Corps, and Air Force.

ABOUT THE AUTHORS

Lawson Magruder
1975 2021

Lawson was honored to be the first company commander of Bravo Company, 2nd Battalion (Ranger), 75th Infantry. He led soldiers in combat in Vietnam and Somalia and as a general officer commanded the Joint Readiness Training Center and Ft Polk (Now Fort Johnson), US Army South in Panama, and the historic 10th Mountain Division. Retiring as a Lieutenant General, he transitioned into the corporate and academic cultures, publicly sharing his leadership journey and serving as an executive coach and mentor for leaders in the federal and private sectors.

His leader development company True Growth, co-owned with Byrd Baggett, helped over 15,000 leaders become more authentic and humble leaders. He is a Distinguished Member of the 75th Ranger Regiment, the US Army Ranger Hall of Fame, the US Army ROTC Hall of Fame, and is a Distinguished Alumnus of the US Army War College.

He coauthored, *True Growth: Simple Insights on How to Live and Lead with Authenticity,* is the author of a personal memoir, *A Soldier's Journey Living His Why.* He has been married to Gloria for over 55 years and they are blessed to have three children and four grandchildren. Lawson's reflection on his service in Bravo Rangers: "It was in Bravo where I learned from my NCOs and peers the true meaning of competence, which is the blending of experience, skill, discipline, and positive attitude."

Fred Kleibacker
1976 2024

Fred Kleibacker (MSG, Ret.) served in 3rd Platoon, B Company, 2nd Ranger Battalion from 1975-1978 as a team leader and squad leader. He reenlisted for Special Forces and served a combined nine years in the 10th and 7th Special Forces Groups, four years in the 1st Special Forces Operational Detachment—Delta, and two years as an ROTC Cadre Instructor at Bucknell University, retiring in 1994. Following retirement, Fred spent eight years as a government civilian employee helping to build a first of its kind, command and staff, Personnel Recovery educational program which trained thousands of soldiers to plan and execute combat rescue missions globally. In 2007, he founded a Service-Disabled Veteran-Owned Small Business providing one-of-a-kind, counter IED, technology to help protect our soldiers for the Defense Department. All of Fred's three children served in the Military: Navy, Army and Coast Guard. Today Fred lives quietly on his homestead with his wife Erika and their two dogs. He remains involved in his local community, is a board member on the Ranger Scholarship Fund, and manages Bravo Company's MG Eldon Bargewell

Memorial Scholarship Award. Fred's quote about his service, "I could not have accomplished what I have in my life had it not been for what I learned in Bravo Company about adaptability, resilience, discipline and teamwork."

www.ingramcontent.com/pod-product-compliance
Lightning Source LLC
Jackson TN
JSHW082100270825
90118JS00017B/219